ENZYME KINETICS AND MECHANISMS

Enzyme Kinetics and Mechanisms

by

KENNETH B. TAYLOR

University of Alabama at Birmingham, U.S.A.

KLUWER ACADEMIC PUBLISHERS
DORDRECHT / BOSTON / LONDON

A C.I.P. Catalogue record for this book is available from the Library of Congress.

ISBN 1-4020-0728-0

Published by Kluwer Academic Publishers,
P.O. Box 17, 3300 AA Dordrecht, The Netherlands.

Sold and distributed in North, Central and South America
by Kluwer Academic Publishers,
101 Philip Drive, Norwell, MA 02061, U.S.A.

In all other countries, sold and distributed
by Kluwer Academic Publishers,
P.O. Box 322, 3300 AH Dordrecht, The Netherlands.

Printed on acid-free paper

Printed in the Netherlands.

To the colleagues who showed the way

and

to the students who taught me.

TABLE OF CONTENTS

PREFACE

It is said that necessity is the mother of invention. Although this seems to be true, I suspect that indolence also has something to do with it and perhaps some paternity testing would be in order. Such was the case with the present book. The necessity was provided by the circumstance that I was to teach a course about a steady-state kinetics approach to the elucidation of enzyme mechanisms, but in the apparent absence of a suitable textbook I was forced to develop my own teaching materials. The putative father of this book was my disinclination to burden myself and the students any more than necessary with long, convoluted derivations and explanations as well as to attempt an exhaustive literature review. Therefore, the approach taken was the shortest and simplest I could find. The success or failure of this approach is to be determined by you the reader.

In addition the ancestry of this book includes the efforts and forbearance of several other individuals. Dr. Tim Fritz, Staff Scientist at the National Institutes of Health, NIDDK, read the manuscript and made many comments that improved the clarity of the book very significantly. His efforts are greatly appreciated.

Dr. W. W. (Mo) Cleland was not only instrumental in the development of much of the science described here but provided me, a number of years ago, with an introduction to it and an opportunity to learn some of it. I am grateful.

Dr. Herb Cheung and Dr. Jeff Engler, both of the Department of Biochemistry and Molecular Genetics at the University of Alabama at Birmingham, provided needed encouragement.

Finally I am grateful to my wife, Carol, who exercised generous forbearance that allowed me to develop the material and write the manuscript during time that otherwise could have been devoted to joint activities and concerns.

Verona, Maine
Summer 2001

INTRODUCTION

Although the rate of appearance in the scientific literature of rigorous investigations of enzyme mechanisms by steady-state kinetics seems to have declined somewhat in recent years, it remains rather steady. Nevertheless the usefulness of such studies remains rather high, because the functions of enzymes are fundamentally kinetic in nature. More specifically the possibilities for rate and mechanistic perturbation due to the substitution of specific amino-acid residues by genetic engineering have enhanced the need for rigorous kinetic studies to define the effects of these changes. With the present availability of site-directed mutants, a plethora of structural information and the possibility for comparison of very closely related enzymes, investigations of steady-state kinetics assume even greater importance.

Two factors further enhance the relevance of investigations of enzyme mechanisms by steady-state kinetics. First the experimental conditions are generally close to physiological conditions. Specifically the enzyme concentration is much less than the substrate concentration both physiologically and for steady-state experiments, as well as necessary for the interpretation of the experimental results. In addition the temperature, the pH, the pressure and the solvent are usually closer to the physiological than is the case with other experimental techniques. Second, the experiments generally require rather simple equipment. As explained in a later chapter a good spectrophotometer and a computer are generally the most sophisticated equipment necessary for most investigations. However, additional analytical and control equipment may be required for some enzymes and for some types of experiments.

The overall objective of the study of enzyme reaction mechanisms is the description of the intermediates and transition states in the reaction. Since there are some intermediates and transition states that are not accessible by these techniques, investigations of steady-state enzyme kinetics have a focus on a subset of these objectives that will be described later.

There are three major objectives of this book. The first is to describe the useful experimental manipulations for the production of steady-state kinetic data as well as their interpretation in order to give the reader an approximation of the magnitude of effort required to complete a meaningful investigation. Second, this book purports to describe the interpretation of data in sufficient detail to enable the reader to understand the principal paths of logic associated with steady-state enzyme kinetics. Specifically the reader should be able to understand the logic in the literature that connects hypothesis and data from steady-state kinetic experiments, and the reader should be able to formulate and use steady-state kinetics experiments and logic in the elucidation of the mechanism of enzymes.

Although it is expected that the reader may want to consult some of the more detailed descriptions of individual approaches that are cited in the text, the present book will be invaluable in the understanding of these descriptions.

The third objective of this book is to expound several approaches to the theory of steady-state enzyme kinetics in a context somewhat different from previous discussions in an attempt to make them relatively easily understood while maintaining a standard of rigorous logic.

This book is not meant to be a scholarly or an exhaustive treatise on enzyme kinetics and mechanisms. Whereas examples from the literature will be presented, a review of the literature is not a purpose of the book. There are a number of excellent and comprehensive treatises on enzyme kinetics and mechanisms to which reference will be made. The present book is intended to aid in the understanding of these treatises. Furthermore, this is not meant to be a reference book, although it may serve that function. There are other compendia of steady-state mechanisms, *e.g.* [1].

This book is for anyone who expends the funds in its purchase. More specifically it is most appropriate for someone who has had an entry-level introduction to enzymes and steady-state kinetics and would like to understand the subject in more detail. The material herein was developed for a graduate level course in enzyme mechanisms and kinetics taught by the author over a period of fifteen years. An understanding of the book will require about an American secondary-school level of knowledge of algebra and analytical geometry. The really necessary concepts will be reviewed very briefly. Some understanding of calculus is necessary to understand the curve-fitting algorithms and the slow inhibitors, but these are not essential for the understanding of steady-state enzyme kinetics. An understanding of the material in this book does require the motivation to indulge in the algebraic thought processes to derive the necessary equations. Equations are central to the approach in this book and are basic for a conceptual understanding of the theory of steady-state enzyme kinetics.

The book is divided into five general sections. In the first section the first three chapters deal respectively with some important basic concepts of steady-state kinetics, methods for the generation of data, and methods for the use of that data in the testing of mathematical models for the mechanism. In the following chapter several methods for the derivation of mathematical models are described. One of these methods will be employed throughout the remainder of the book. Therefore, it is important to have an understanding of this chapter in order to understand the remainder of the book. The third section consists of four chapters in three of which models are described respectively to deal with initial-velocity data from experiments in which the concentration of substrate, of analog inhibitors, and of product inhibitors are varied respectively. The latter section also includes a chapter about models for substrate inhibition. In a departure from strictly initial velocity models the fourth section consists of a chapter containing a description of models for tight-binding inhibitors, slow-binding inhibitors and slow-, tight-binding inhibitors, because of their importance and because the models depend on many of the same concepts as initial-velocity models. The final section contains Chapter 10 with a discussion of the thermodynamics of initial velocity and general models for the description of the effects of changes in environmental conditions, and other reaction conditions on the initial velocity. An understanding of this chapter is important for the understanding of the final three chapters, which present specific models for the description of the effects of pH, isotopic substitution and other factors, such as temperature and pressure, on the initial velocity.

References

1. Segal, I.H. *Enzyme Kinetics*, Wiley Interscience, New York, 957pp (1975).

CHAPTER 1

STEADY-STATE KINETICS

1.1. Introduction

This book contains a number of concepts and agendas implicit in steady-state enzyme kinetics that will hopefully be made explicit in this chapter. Specifically this chapter will contain a description of the steady state as well as the assumptions and approximations associated with it. It will contain a general description of the things that can be learned from and the limitations of investigations of steady-state enzyme kinetics. Finally a generally useful sequence of experiments in an investigation of steady-state enzyme kinetics will be described.

1.2. What is the Steady-State?

In a prototypical experiment an enzyme reaction is initiated by the combination of free enzyme and substrate rather instantly compared to the other things that happen, whether this is accomplished by the simple addition of enzyme to a reaction with a pipette or it is done with a very fast stopped-flow instrument. The period immediately after the initiation is characterized by the increase in concentration of the downstream intermediates of the reaction and is called the pre-steady-state period. The pre-steady-state period is followed by a second period during which these intermediates of the reaction are in relatively constant concentration. During the latter period the approximations necessary for steady-state kinetics are most accurately realized, and it is called the steady-state period. During the steady-state period the rate of the appearance of product is most nearly constant. This is called the initial velocity (v_i) of the reaction.

These phenomena can be illustrated in a simulated reaction (Figure 1.1) in which the free enzyme (E) forms an enzyme-substrate intermediate (EA) and the latter subsequently forms an enzyme-product intermediate (EP), which dissociates to product (P) and free enzyme. It can be seen that the product concentration endures a short lag period, while the concentration of both intermediates increases with time. Then the product concentration increases rather linearly with time, while the concentration of both intermediates remains relatively constant. This second period is the steady-state period and the rate of formation of product is the initial velocity. The previous lag period is frequently

1

sufficiently short that it is insignificant in the measurement of initial velocity. In fact special equipment is frequently required to investigate this pre-steady-state period. Finally the product concentration will increase more slowly as the reaction either approaches equilibrium or the substrate concentration becomes so low that the rate of formation of the enzyme-substrate complex (EA) becomes rate limiting, or both. In

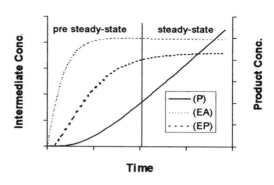

an irreversible reaction only the rate of formation of the enzyme-substrate complex limits the reaction rate in this third period.

1.3. Assumptions and Approximations

There are five interrelated assumptions that must be at least approximated in order to perform and interpret valid initial-velocity experiments (Table 1.1)

Table 1.1. Assumptions for initial-velocity experiments.

1.   The concentration of all enzyme forms are constant during the measurement.

2.   The measured initial velocity is the steady-state velocity.

3.   The concentration of substrate (and inhibitors) is constant.

4.   The concentration of at least one product in negligible during measurement.

5.   All other reaction conditions are constant during measurement.

Although not strictly true the assumption of the constant concentration of all enzyme forms in the reaction is most closely approximated during the steady state. The interpretation of data and the mathematical models are predicated on this assumption, and the other assumptions are related to it.

It is assumed that initial velocity can be measured or at least approximated closely, and that this is a measure of the velocity during the steady state. Ideally, but not always, this rate is constant within the limits of measurement precision for a finite period of time, and this constancy is a criterion for initial velocity. For reasons that should be investigated some enzymes under some conditions demonstrate a burst or a lag in the velocity at the beginning of the reaction, which may or may not reflect the pre-steady-state rate. It is not unusual for an investigator to ignore the burst or lag and measure the constant rate that follows.

The steady-state period is prolonged when the concentration of substrate is much greater than the concentration of enzyme, perhaps at least 100-fold. In addition as demonstrated later this is a requirement and simplifies the derivation of mathematical models, since it circumvents the necessity for the acknowledgment of the disappearance of substrate in derivation of steady-state models. Therefore, it is assumed that the concentration of substrate remains the same as that put into the reaction mixture initially. This requirement for a relatively small concentration of enzyme sometimes limits the kinds of enzymes with which steady-state investigations can be conducted. For example investigations with polymerases and nucleases are limited because an easily detectable reaction rate frequently requires concentrations of enzyme comparable to that of the polymer template or substrate respectively.

In addition to the constant substrate concentration it is assumed that the concentration of the product; or at least of one of the products, if there are more than one; of the reaction is zero during this period. This approximation also simplifies the derivation of mathematical models considerably, since the step in which this product is released is considered to be irreversible, and there is no overall reverse reaction with which to be concerned. This approximation sometimes seems to be a contradiction in terms, since one may be measuring the rate of formation of product before there is significant product. However, the apparent contradiction is the same as that associated with any instantaneous rate or slope and the approximation does not generally present practical problems.

The final assumption is that the change in other conditions during the period of measurement of initial velocity is negligible. Specifically it is assumed that the pH, temperature, and ionic strength are constant during the period of measurement. Although the pH is generally controlled by the use of a buffer, some precautions will be discussed later along with other methods. Temperature control is a technical matter and will also be discussed along with other methods. Changes in ionic strength are generally not a problem.

Operationally there are some tests that may lend confidence that initial velocity is actually measured, but these will be discussed along with other methods.

1.4. What Can be Learned?

In general several kinds of mechanistic hypotheses can be tested with data from steady-state kinetic investigations. Most commonly one can learn about the substrate binding steps, the steady-state mechanism. It is possible to test hypotheses about ordered binding, random binding, rapid-equilibrium binding and other similar hypotheses. In addition it is possible to test hypotheses about the order of product release. When the initial velocity can be measured in both directions, one can learn about the substrate binding steps in both directions. Furthermore in favorable cases minor pathways of substrate binding and product release can be identified.

Steady-state data can be used to test certain kinds of hypotheses about the steps between substrate binding and product release and more limited hypotheses about the steps between product release and substrate binding. The meaningful data comes from experiments in which the steady-state kinetic parameters are measured under various environmental permutations (*e.g.* pH, temperature, ionic strength), enzyme permutations (*e.g.* site-directed mutants) and substrate permutations (*e.g.* analogs, isotopic substitution). Hypotheses can be tested about the identification of the rate-determining processes and the extent to which certain processes are rate-determining can be determined. For example it can be determined whether the rate-determining processes involve the breaking, the forming or changes in hybridization of specific bonds in the substrate. In addition essential acidic or basic groups on the enzyme can be detected. Furthermore, hypotheses about their identity as well as their role in the reaction can be tested.

1.5. The Limitations

In spite of the optimistic possibilities for steady-state data there are a number of limitations to the kinds of enzymes that can be investigated and the hypotheses that can be tested. Therefore, the elucidation of an enzyme mechanism requires the steady-state data and the hypotheses confirmed or eliminated along with data and hypotheses from other techniques such as x-ray crystallography, fast-reaction kinetics, and others.

Certain kinds of enzymes are difficult to study. Polymerase and depolymerase enzymes are frequently difficult, because the polymer, either template or substrate, is frequently necessary in concentration comparable to that of the enzyme in order to measure a significant reaction rate. Therefore, the polymer cannot be treated as a substrate for steady-state modeling purposes, and investigations with these enzymes are limited.

In addition it is possible to test hypotheses about the number and kinds of intermediates and transition states between the binding of substrate and the release of product only indirectly by the effects of certain reaction conditions on the steady-state parameters. Furthermore the testing of hypotheses about specific chemical groups involved in the catalytic reaction is limited to the possible identification of specific essential acidic and basic groups on the enzyme under favorable circumstances.

Finally there are technical reasons why particular enzymes cannot be investigated as thoroughly as desired by steady-state kinetics. The most common reason is the absence of a satisfactory method by which to measure initial velocity. For reasons explained later it is very advantageous to be able to measure the initial velocity continuously in real time. However, the investigator's ingenuity is sometimes challenged beyond its limits to find a satisfactory method to accomplish such measurements and less satisfactory methods must be employed.

Another common reason is the relative scarcity of the physiological substrate in a pure or even a well characterized preparation. For example investigations of a number of the hydrolases are limited by the fact that the physiological substrate is relatively scarce and expensive. In many cases even if it were readily available, neither it nor the product has a unique physical property by which to detect its disappearance or appearance respectively in a continuous manner in real time.

1.6. A Sequence of Investigation

Although they are not always conducted in this sequence, there is a favored logical sequence to steady-state investigations. Ideally an enzyme should be investigated in this order. Some detailed planning of the investigation at the beginning can prevent the laborious repetition of some experiments later. For example there is some advantage in the selection of a pH, temperature and substrate with which the reaction can be investigated with some convenience in both directions. However, the results and conclusions will be more useful if these conditions are kept as close to physiological as possible.

It is useful to spend some time at the beginning to establish the reaction conditions under which the remainder of the investigation will be carried out. The method for the measurement of initial velocity should be rapid and convenient and result in precise, repeatable data. It is best to design a method that is sufficiently robust that equipment is not operating at its limits in order to avoid instrumental variability later. The conditions, particularly the pH, should be easily controlled and such that the initial velocity is relatively insensitive to small changes. It is also useful to streamline the manipulations required in the assay procedure, because it will be repeated many times.

After preliminary experiments are completed to establish the best conditions for the rest of the investigation, the investigation can be conducted in approximately the same order as the chapters in this book. First it is useful to generate the data and test hypotheses about the steady-state mechanism for binding of substrate and product. The latter generally requires data from experiments in which the initial velocity is measured at various concentrations of substrate, reversible analog inhibitors and product inhibitors. It is useful to carry out the experiments in both directions of the reaction, if possible. This is also a good time to test hypotheses about the release of product in those cases in which it is possible. Next, it is common to perform the experiments and test the hypotheses to establish minor binding pathways, if possible. The latter requires data from experiments with substrate inhibition.

Next it may be useful to perform the experiments to test hypotheses about the effects of pH on the initial velocity, and finally is is useful to investigate isotope effects as well as some of the environmental effects in order to test hypotheses about the rate-determining steps. Data from experiments to measure the effects of changes in reaction conditions is more easily interpreted if hypotheses about the steady-state mechanism have previously been verified. However, since it is possible to investigate isotope effects as well as the effects of other substrate permutations without actually measuring the initial velocity *per se*, some of the initial-velocity investigations may be attenuated somewhat in order to hasten the investigation of the former, particularly in those cases in which the initial-velocity measurements are difficult, relatively unreliable or both.

The remainder of this book is to supply some of the methods and the logic by which these investigations might be carried out.

CHAPTER 2

THE GENERATION OF EXPERIMENTAL DATA

In the present chapter will be presented some of the methods for the generation of initial-velocity data. It is intended to give the reader a conceptual picture of the nature of the investigation rather than to provide a comprehensive or complete description of every method. In addition attempts are made to describe procedures that have been particularly useful to improve the precision of the data and to minimize the manual labor involved. Finally methods are described for the calculation of initial velocity from the data output of the instruments used.

Most frequently the investigator will measure the initial velocity at various concentrations of substrate and frequently at different values of at least one additional reaction condition, which may be the concentration of an additional necessary substrate, the concentration of a reversible inhibitor, pH, or the isotopic composition of some atom of the substrate.

2.1. Experimental Objective

The objective of the experiments is to measure the concentration of product or of substrate as a function of time elapsed, subsequent to the initiation of the reaction, and to determine the slope of the initial part of the curve relating product concentration and time (*e.g.* Figure 1.1) or the corresponding initial part of a substrate-concentration curve. It is usually more precise to measure changes in the product concentration than in the substrate concentration, since the former will be a small change in a small concentration rather than a small change in a large concentration. However, frequently other considerations take precedence. Of course there are many methods for the determine the concentration of product or substrate, but these will be a subject of the present chapter only in a parenthetical way.

2.2. Experimental Methods

There are three general methods by which the initial-velocity data is generated, continuous, discontinuous and coupled methods.

7

2.2.1. CONTINUOUS METHODS

If either the substrate or the product of the reaction has some unique physical property that can be measured in real time under the conditions of the reaction, that property may be monitored continuously as a measure of the reaction progress. Most commonly advantage is taken of the selective absorption of light by either a product or a substrate. Alternatively, however, advantage may be taken of a selective change in another physical property. For example the product of alcohol dehdrogenase (Figure 2.1), NADH, absorbs light at 340 mμ. The success of this method is evident in the many steady-state kinetic investigations of dehydrogenases. The continuous generation of data with analog recording or the generation of numerous data points with digital recording permits a very precise measurement of initial velocity.

$$NAD + RH_2COH \rightleftharpoons RH_2CO + NADH + H^+$$

2.2.2. DISCONTINUOUS METHODS

Figure 2.1

When neither the substrate nor the product can be selectively measured under the conditions of the reaction, the reaction must be stopped and the measurement carried out under different conditions. Measurement usually requires the chemical separation of substrate and product with quantitation of one of them. A common example is the use of a radioactive substrate with subsequent separation of substrate and product and the determination of the radioactivity associated with the latter. This generation of a single time point with each analytical sample constitutes a discontinuous measurement method.

Although frequently analysis at a single time point will provide an adequate measure of initial velocity, initially a time course consisting of a number of samples should be analyzed in order to establish that initial velocity is being measured. When the reaction conditions are changed substantially this time course should be repeated. It can be seen that the manual labor involved in a significant kinetic investigation by these methods can become substantial. For this reason and because of the additional approximations involved, discontinuous methods are less preferable that the continuous ones. If a single analysis by a discontinuous method requires a significant amount of time, it is probably useful to spend time working out some sort of continuous method before the initiation of a steady-state kinetic investigation.

2.2.3. COUPLED METHODS

In some cases a discontinuous assay method can be avoided, when a product of the reaction of interest can be the substrate of a second enzyme whose reaction can be measured continuously. This constitutes a coupled assay method.

The second enzyme, the coupling enzyme, as well as any necessary additional substrates should ideally be readily available in quantity, since it is necessary to employ them in high concentration in order to insure that the coupling reaction keeps up with the initial velocity of the reaction under investigation. Experiments should be done with various amounts of the components of the second enzyme reaction in order to demonstrate that initial velocity is being measured, and these experiments should be repeated whenever the reaction conditions are changed significantly. Although there is a mathematical model with which the adequacy of these components can be calculated [1], it is done more easily by measurement of the initial velocity with twice as much of the components for the coupling reaction and determination whether the apparent initial velocity has increased significantly. If not, there were probably enough of the components of the second reaction at the lower concentration. If it does increase, additional determinations must be done to insure that enough is present.

Since the coupled assay methods are continuous methods, considerable ingenuity has been expended in their development and it is common to utilize more than one coupling enzyme. For example the use of pyruvate kinase (PK) and lactate dehydrogenase (LDH) in the presence of phosphoenolpyruvate (PEP) and NADH is a common coupling system for many phosphotransferase enzymes (t'ase) that produce ADP as one of their products (Figure 2.2). With this method the disappearance of NADH is monitored spectropotometrically at 340 mμ. The method has the additional advantage that the ADP is recycled and the phosphotransferase reaction is rendered essentially irreversible.

$$R_2HCOH + ATP \overset{t'ase}{\rightleftharpoons} R_2HCOP + ADP$$

$$ADP + PEP \overset{PK}{\rightleftharpoons} pyruvate + ATP$$

$$NADH + pyruvate \overset{LDH}{\rightleftharpoons} lactate + NAD$$

Figure 2.2

2.3. Reaction Conditions

Of course those conditions that affect the rate of the reaction (e.g. temperature, pH and ionic strength) must be as constant as possible both during a single measurement run and from one run to another. It is common to monitor the reaction for a period of time prior to initiation in order to confirm that the conditions are either constant or not affecting the measurements being made.

2.3.1. TEMPERATURE

The temperature is generally controlled (e.g. ± 0.1 °C.) by the use of a thermostated bath or chamber in which the reaction takes place. Although the reaction temperature can be the ambient temperature, the use of a thermostated bath or chamber is nevertheless recommended, since the temperature in a room may change significantly from one day or hour to the next. In addition a reaction temperature ten to fifteen degrees above ambient can be controlled precisely and requires only heating for good control. Therefore, 37 °C is a good reaction temperature in the interest of both good control and the physiology of many of the enzymes of interest.

Also in the interest of good temperature control the reaction mixtures are equilibrated to the reaction temperature before initiation of the reaction. It is usually best to use a water bath for this purpose, since the heat conductivity is more rapid in it than in air. Frequently the reaction is initiated by the addition of a solution of enzyme that has been maintained at 0 °C to 4 °C in the interest of stability. In order to maximize the temperature control in these circumstances the enzyme, or any other initiating agent, should be transferred in a sufficiently small volume that the temperature of the reaction mixture is not affected significantly. Alternatively the initiating component can be pre-equilibrated to the reaction temperature.

Temperature control at temperatures farther than twenty degrees from ambient is complicated by the fact that it is difficult to make transfers of significant amounts of solution to the reaction container with the maintenance of strict temperature control.

2.3.2. CONTROL OF pH

The pH of the reaction mixture is generally the optimum pH of the enzyme, the physiological pH of the enzyme, or both, if they coincide. However, it may be quite different from either of these for special purposes. Nevertheless the initial-velocity assumptions and approximations should be verified whenever there is a significant change in the pH of the reaction. The pH control during the reaction is generally maintained by the use of a buffer. Nevertheless the investigator must be sensitive to the need to have a sufficient concentration of buffer at a pH at which it has adequate buffering capacity. In addition the measurement of the pH of at least one reaction mixture before and after the reaction to confirm that the pH is accurate and that there is no significant change during the measured reaction will provide a degree of confidence in the effectiveness of the buffer.

2.3.3. IONIC STRENGTH

The ionic strength of most biological reactions changes rather little during the reaction, although the investigator should remain sensitive to the possibility. However, the addition of ionic substances to subsequent reaction mixtures, particularly inhibitors, may change the ionic strength significantly. When such a possible change is an issue, it may be useful to determine the effect of a change in ionic strength more selectively by the addition of salt in comparable concentration to the reaction mixture.

2.3.4. SUBSTRATE CONCENTRATION

In the experimental protocol generally it is good to have at least five different concentrations of substrate in each series of measurements. Unless others factors take precedence, *e.g.* cost or solubility, a good rule of thumb is to have the range of concentrations of substrate extend from one-fifth of the K_M to five times the K_M. In order to promote uniformity of all of the reaction mixtures in an experimental series combine as many of the nonvariable components of the reaction mixture as possible in a stock solution. After the stock solution has been distributed into a series of separate reaction mixtures, add the variable component, usually substrate.

In addition it is generally necessary to determine the exact concentration of substrate in the stock solution of substrate in spite of the fact that it was carefully weighed and made up to volume. It is frequently possible to conduct this determination enzymatically in the presence of larger amounts of the same enzyme and sometimes under somewhat different conditions (*e.g.* pH) in order to encourage the reaction to go to completion and do so in a convenient period of time.

Although reaction mixtures of very small volume (*e.g.* 0.1-0.2 ml) are to be avoided, if such volumes are necessary, special care must be taken to insure or at least measure the precision of the volumetric transfers. Generally a microliter syringe is preferable to a pipettor with changeable tips. The latter is more acceptable, if it is very carefully calibrated, maintained and used. In addition to volumetric transfers special care should be taken to insure complete mixing of such small volumes.

2.3.5. INITIATION OF THE REACTION

Frequently the reaction is initiated by the addition of a rather small volume of enzyme or possibly substrate. The amount added must be as precise as possible. It is useful to calibrate the instrument used for the addition in order to estimate its precision. The addition of amounts less than about ten microliters and the calibration of the corresponding instrument are both difficult to accomplish with acceptable precision. In addition, whether it is the initiation agent or not, care must be taken to insure that the same amount of enzyme is used from one experimental session to another, particularly when they are separated by a significant period of time. It is frequently necessary to assay the enzyme activity carefully before each experimental session to insure its uniformity.

The addition of enzyme, or substrate, and its even distribution in the reaction mixture to initiate the reaction should be done rather quickly in order to insure the collection of data constituting the initial velocity. For example in spectrophotometric measurement methods it is possible to fabricate or purchase a stirring device consisting of a small rod with a larger spoon or block on one end that transfers a small amount of liquid and fits inside the cuvette. A few vertical motions with this device will mix the solution in the cuvette.

2.4. Calculation of Initial Velocity

There are a number of ways in which to calculate initial velocity from data, depending on the method and the form of the data. In continuous measurement methods the initial velocity can be determined from graphic data from a chart recorder by the construction of the tangent with a straight edge and calculation of the slope of the tangent. Tabular data in a computer file in a spreadsheet can be fit with a polynomial equation, usually second or third order (methods in Chapter 3). The coefficient of the first-order term is the initial velocity, since it is the value of the first derivative at the origin.

Data from discontinuous measurements frequently consists of a single point at a time predetermined to be within a satisfactory approximation of the initial velocity period. As stated above it is advisable to confirm this approximation frequently.

However, every method is associated with some dead time at the beginning of the reaction during which acceptable data cannot be gathered. The dead time, which is most apparent in continuous measurement methods, can be dealt with in several ways. If it is quite short, it is usually ignored and the subsequent rate is measured as a sufficiently close approximation of the initial velocity. If the dead time seems significant or if there is reason to doubt the validity of the subsequent rate, the data can be extrapolated to zero time by a fit to the data of a polynomial equation as described above or of the integrated Michaelis-Menten equation. In addition the extrapolation to zero time is sometimes employed to determine the initial velocity in the presence of an unexplained lag or burst of initial activity.

When possible the initial-velocity data should be verified by testing it for linearity. The pre-steady-state portion of the initial concentration changes is usually insignificant with the usual methods for measurement. However, data should be inspected and tested for confirmation of this fact. This can be accomplished by testing graphical data with a simple straightedge or more elegantly with tabular data in a spreadsheet by linear regression. If the initial-velocity period is so short that it must be determined by differentiation of the progress curve, either by graphical or analytical means; it might be verified by demonstration that the value is proportional to the concentration of enzyme.

2.5. Data Handling

Detailed data handling including the determination of the goodness of fit to mathematical models, including the Michaelis Menten model, as well as the determination of the parameter values by computer program is discussed in the following chapter. However, it is good to plot the data initially, even while subsequent data collection is under way, by one of the linear graphical methods such as double reciprocal, Lineweaver-Burke, plots in order to provide assurance that there are no large problems with the data, while there remains an opportunity to correct them.

2.6. References

1. Easterby, J.S. "Coupled Enzyme Assays: A General Expression for the Transient," *Biochim. Biophys. Acta*, 293, 552-8 (1973).

CHAPTER 3

METHODS FOR MODEL EVALUATION

3.1. Introduction

In investigations of steady-state enzyme kinetics generally two kinds of related questions are asked. First, which mathematical model best describes the data? Second, how do the kinetic parameters change with specific changes in reaction conditions, and is the change significant? In the approach to the first question the data generated by the methods outlined in the previous chapter will be tested for its fit to various mathematical models, or equations. The values of the kinetic parameters result from the fit of the data to the appropriate mathematical model. Thus the second question is approached when these values are compared under different conditions.

The development of the models and their testing with data is known as mathematical modeling. The present chapter describes the use of mathematical modeling in testing models for data from investigations of steady-state enzyme kinetics. The purpose here is for the reader to achieve an understanding of the process as well as the assumptions involved and to be comfortable with the software available. Future chapters will deal with the development of models. The reader is referred to more specialized descriptions for additional details [1],[2].

Most often the various models will be fit to data by computer program. The remainder of this chapter is meant to acquaint the reader with the algorithms that make up the programs and with some of the specific programs themselves. However, in subsequent chapters of this book the various models will be discussed in the context of graphical presentation as double reciprocal plots, *i.e.* Lineweaver-Burke plots.

3.2. Mathematical Modeling

A mathematical model is an equation that contains one dependent variable, ordinarily the initial velocity; one or more independent variables, *e.g.* substrate concentration; and one or more parameters, *e.g.* the Michaelis constant. Unfortunately there is no way to test whether a given model is the correct model. It is only possible to compare one model with another. Therefore, it is only possible to identify the best model in a set. It is even more convincing if the members of the set are closely related. Thus the selection of a model is indirect, and the models in a set that do not fit very well are at least as important as the one that fits best.

As will be seen in the remainder of the book the marriage of steady-state enzyme kinetics and mathematical modeling has been a productive and happy one. One of the reasons for the rather happy marriage between steady-state enzyme kinetics and mathematical modeling is that the investigator is frequently faced with data and only a limited set of reasonable mathematical models that are also related to each other. Furthermore, in the author's experience the selection of a best model from a small set of chemically possible models to fit data from steady-state experiments has not generally been a difficult choice.

Nevertheless, the investigator must be sensitive to the fact that in reality many enzymes may have a chemical mechanism intermediate between two models. For example an enzyme may result in data that fits a model for the sequential ordered binding of substrates, whereas there may actually be slow steps for the opposite order of binding. Thus, although the mechanism is partially random, the methods described may not be sufficiently sensitive to detect it.

3.3. Data Fitting Methods

Before several mathematical models can be compared, the best fit of the experimental data to each of the several models must be achieved. Operationally this process is to find the optimum values for the parameters in the mathematical model that will minimize the difference between the experimental values and the calculated values of the dependent variable, *i.e.* initial velocity. For example if the equation is the single-substrate Michaelis-Menten equation (equation 3.1, derived in the following chapter), the process is to find the optimum values of K_M and V_{max} so that the values of initial velocity calculated at each experimental substrate concentration are as close as possible to the values actually measured.

$$v_i = \frac{V_{max} *(S)}{K_M + (S)}$$

(3.1)

Traditionally this has been done graphically by transformation of the model equations to give an expression that is linear in the experimental variables, *e.g.* the double reciprocal plot (Lineweaver-Burke plot) and related ones. The straight line is drawn either graphically or mathematically by linear regression. The problem with these methods is that the transformation of the data yields a new dependent variable whose values have different standard deviations from each other, whereas regression or even graphic line construction rests on the assumption that all points have the same standard deviations. For example if the standard deviation of the original measurements of initial velocity is $\sigma(v)$, and the new dependent variable, y, is $1/v$; the standard deviation of y, $\sigma(y)$, will depend on the value of the initial velocity itself (equation 3.2). The smaller values of initial velocity will have a greater standard deviation than the larger values [3].

$$y = \frac{1}{v_i}$$

$$\sigma(y) = \frac{\sigma(v)}{(v_i)^2}$$

(3.2)

Although there are methods by which this weight can be compensated, it is more rigorous statistically and more simple to fit models to data with no transformation of the data. Therefore, in steady-state kinetics it is more satisfactory to fit data; *e.g.* initial velocity and substrate concentration with no operations or transformations of the experimental values [4]. Ways will now be examined to optimize the parameter values in the equations without transformation of the data.

It is generally surprising how quickly a visually good-looking agreement of a mathematical model with data can be achieved by repetitive manual adjustment of the values of the model parameters (*e.g.* K_M and V_{max}) in a spread sheet program constructed to plot the data and the calculated curve on the same graph, the cut-and-try method. However, statistically rigorous methods must be used eventually to be as certain as possible that the best fit was achieved, to estimate the goodness-of-fit, and to estimate the degree of confidence in the parameter values. Therefore, we must consider three questions: What is the best measure of how well a given curve fits the data (goodness-of-fit)? What is the best way to find the optimum values for the parameters? What is the best way to estimate the degree of confidence in the resulting optimum values?

3.3.1. THE MEASURE OF GOODNESS-OF-FIT

The least-squares function, s^2, is the sum of the squares of the differences between the experimental values, v_{exp}, and the calculated values, v_{calc}, of the dependent variable (equation 3.3, Figure 3.1). It is by far the most widely used and most generally accepted measurement of goodness-of-fit.

$$s^2 = \sum \left(r^2 \right)$$

$$s^2 = \sum \left(v_{exp} - v_{calc} \right)^2$$

$$s^2 = \sum \left(v_{exp} - \frac{(A_{exp}) * V_{max}}{(A_{exp}) + K_m} \right)^2$$

(3.3)

However, its validity requires some assumptions. Other methods are available for cases in which these assumptions cannot be approximated.

The most important assumptions are: 1)All of the significant error occurs in the dependent variable (initial velocity in our case). This is not too difficult for our purposes, if care is taken with pipeting so that the error in the concentrations of substrates, inhibitors, *etc.* are minimal. 2)All of the data points are equally precise, *i.e.* have the same standard deviation. If there is an estimate of the precision of each of the data points separately, violations of this assumption can be compensated for by weighting.

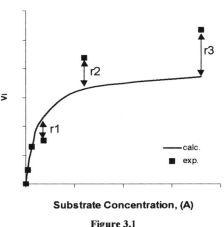

Substrate Concentration, (A)

Figure 3.1

3)All of the significant error is random, and there are no systematic errors. This assumption should receive some thought in each study, since systematic errors can frequently either be identified and eliminated by revision of the experiment or corrected for by weighting of the data. In addition some physical measurements may contain some significant but subtle nonrandom error. 4)There are sufficient data points to provide a valid sample of the experimental results. Although there must be at least as many data points as unknown parameters in the model equation, ordinarily considerably more are required. This requirement was discussed in the previous chapter. 5)The values measured for the dependent variable are independent of each other, or at least there are not relationships among them that are not accounted for in the model.

3.3.2. SEARCH METHODS FOR OPTIMUM PARAMETER VALUES

Several algorithms for the fitting of data to mathematical models are available. Although a completely rigorous description of the methods is beyond the scope of this book, it is desirable to provide enough insight into some of the various methods so that the reader might feel comfortable using them. The reader is referred to other sources [1] for a more detailed discussion.

The objective of the algorithm is to determine the best (optimum) parameter values so that the calculated values of the dependent variable (usually initial velocity) are as close as possible to the experimental values. It is to find the parameter values to give the minimum value of the least-squares function. Operationally the algorithm will set up a series of simultaneous, linear equations that can be solved for the parameter values. This process of parameter optimization is known as curve fitting.

Models with Linear Parameters
The algorithm is considerably simpler if the model equation is already linear with respect
to the parameters to be estimated, because the first derivative of the least-squares function
will also be linear. Thus the parameters must be in the numerator, must be no higher than
first power, and appear in no cross products with other parameters. For equations that are
linear in each of the parameters to be optimized the first derivative of the least-squares
function is determined with respect to each of the parameters. Therefore, the number of
resulting equations equals the number of parameters. The first derivative of the original
least-squares function is the slope of that function, which will equal zero at the minimum
of the function. Therefore, each equation is set equal to zero and the system of
simultaneous linear equations is solved analytically by whatever method you choose (*e.g.*
Cramer's rules, matrix inversion, *etc.*). For example data can be fit to the polynomial,
equation 3.4, because it is linear with respect to the parameters to be estimated, a, b, and
c. For details see Appendix 3.1, section 3.9.

$$y = a * x^2 + b * x + c \qquad (3.4)$$

Models with Nonlinear Parameters
However, if any of the parameters in the model equation is nonlinear, the first derivative
of the corresponding least-squares function will be nonlinear and the system of equations
will be impossible to solve analytically. Unfortunately the Michaelis-Menten and similar
equations are nonlinear in K_M (third line, Equation 3.3). The algorithms for equations that
are nonlinear in any parameters are done by reiterative, numeric methods, in which a
convergent series of parameter estimates produces subsequently smaller values for the least-
squares function. Most methods require the input of the mathematical equation, the first
derivatives of the equation with respect to each of the parameters, and initial estimates for
the values of each of the kinetic parameters. The algorithm then refines the initial
estimates to find the best values for the parameters.

The reiteration will converge much more quickly and surely, if the initial
parameter estimates are close to the optimum values. If the initial estimates are too far
from the optimum, the iteration will diverge and the computer program will crash or will
yield unreasonable parameter values. Therefore, good estimates are important. The initial
estimates for the Michaelis-Menten and similar equations are frequently done by a linear
transformation, *e.g.* double-reciprocal equation. Alternatively the initial estimates can
usually be done rather quickly by the "cut-and-try" method described above.

Since most available computer programs determine the first derivatives
numerically, analytic equations for the first derivatives are usually not necessary.

The concept of error space is useful in discussions of nonlinear curvefitting. It is
a graphical coordinate system in which the value of the least-squares function is
represented in one dimension, usually vertical, and the value of each of the parameters is

represented in the other dimensions. Therefore, a model equation with "n" parameters would have n +1 dimensions. For example the Michaelis-Menten equation with two parameters (V_{max} and K_M) would have three dimensions. Graphically the value of the least-squares function is represented by a surface and the point where it is a minimum on the surface is the optimum value of each of the parameters. Although it has not been a common feature associated with these efforts, it must be realized that the algorithm may find false minima on this surface. The probability of false minima can be reduced by the conduct of searches from more than one set of initial parameter estimates.

3.4. Methods for Parameter Optimization

Four methods for parameter optimization will be described in an attempt to give the reader an intuitive understanding of each. They are: the line-of-steepest-descent, the Gauss-Newton method, the Levenberg-Marquardt method, and the simplex method.

3.4.1. LINE-OF-STEEPEST-DESCENT METHOD

The value of each parameter estimate is changed in the direction in which the change produces the greatest decrease in the least-squares function. Thus, if the derivative of the least-squares function with respect to V_{max} is positive and twice that with respect to K_M, the next estimate will decrease the V_{max} twice as much as K_M. Exactly how far the next estimate is changed along this vector varies somewhat from one program to another. This method converges rather rapidly for estimates that are far (but not too far) from the optimum, but it is rather slow for estimates that are close to the optimum. Therefore, it is frequently used for introductory refinements of the initial estimates.

3.4.2. GAUSS-NEWTON METHOD

The model equation is approximated by Newton's reiterative method, found in most calculus books, of solving an equation, *e.g.* $\int(x)$. It is more formally expressed as a Taylor series truncated after the first derivative (Equation 3.5).

$$y = f(x_1) + (x_2 - x_1) * f'(x_1) \qquad (3.5)$$

Thus the value of the independent variable, x, to give a specified value to the dependent variable, y, can be determined in a reiterative manner. According to equation 3.5, y equals the sum of its value at some value of x, x_1, estimated to approximate the desired value of y and the first derivative of the function at the estimated value of x times the difference between the estimated value of x, x_1, and an improved value of x, x_2. The equation is then solved for the latter difference and the improved value of the x, x_2, is calculated for use in the next round of iteration. For example if the roots of the equation are desired, set the value of y equal to zero, estimate the best value for x_1, and solve for the difference between x_1 and x_2. Find the value of x_2 and start the iteration over. Repeat the process until the differences become very small.

This method was first applied to the Michaelis-Menten equation by Wilkinson [3], and a more systematic explanation of it is provided by Cleland [5]. Application to the Michaelis-Menten equation looks like equation 3.6 where the first term is the Michaelis-Menten equation evaluated with the first estimates of K_M and V_{max} (K_{M0} and V_{max0}), the second term is the product of the first derivative of the Michaelis-Menten equation with respect to K_M and the difference the first estimate and the second estimate of K_M; the third term is the product of the first derivative of the Michaelis-Menten equation with respect to V_{max} and the difference of the first estimate and the second estimate of V_{max}[1].

$$v_i = \frac{V_{max0}*(A)}{(A)+K_{M0}} + (K_{M1}-K_{M0})*\frac{d}{dK_M}\left[\frac{V_{max0}*(A)}{(A)+K_{M0}}\right] + (V_{max1}-V_{max0})*\frac{d}{dV_{max}}\left[\frac{V_{max0}*(A)}{(A)+K_{M0}}\right] \qquad (3.6)$$

Since this equation is linear in the parameter differences (e.g. K_{M1}- K_{M0}), the corresponding least-squares function can be fit to the data analytically as described above and the resulting parameter values are the differences (e.g. K_{M1}-K_{M0}). From the difference of the parameter values and the first estimate of them the next value for the parameter estimate can be calculated, tested for goodness-of-fit, and then used to start the next iteration. This loop is repeated until both the parameter estimates and the least-squares function change less than some predetermined small fraction, which can usually be set by the operator of the program.

The Gauss-Newton method is better for refinements of the estimates later in the iteration process, since it sometimes produces a divergent sequence of the least-squares function, if the parameter estimates are not rather accurate.

---

[1] Equation 3.5 is actually more complicated than necessary since the Michaelis-Menten equation is actually linear in V_{max}. However, the underlying concepts are easier to understand as written and the algorithm works as written.

3.4.3. MARQUARDT-LEVENBERG METHOD

The Marquardt-Levenberg method is a combination of the previous two methods in which initial optimization is done with the line-of-steepest-descent method and the later optimization is done by the Gauss-Newton method. There are a number of variations of this method particularly in the algorithm for the change from one method to the other for optimization. The direction of parameter change for each of the two methods can be regarded as a vector. Different computer programs will use a combination of the two vectors in various proportions depending on the number of iterations and the degree of previous convergence.

3.4.4. SIMPLEX METHOD

This is conceptually a geometric method, although it is carried out mathematically. It is used by some of the commercial programs for primary refinement of initial estimates of parameter values prior to more definitive optimization. It has the disadvantage that the value of the least-squares function converges rather slowly but has the advantage that it will not diverge. In addition this method does not require first derivatives of the model function with respect to each of the parameters.

Very briefly it expands, by random number generation, the number of sets of initial parameter estimates from one to $n + 1$, where n is the number of parameters. It then tries new parameter values in the opposite direction of the worst estimate relative to a centroid formed by the other estimates in error space. If the error value (least-squares function) at the next estimate set is less, it may keep them, discard the worst estimate and start the process over or even proceed further in the same direction. Otherwise it may try a parameter set less far in the original direction or even start over with new parameter estimates.

For example with the Michaelis Menten equation it would be searching for two parameters and the error space would be in three dimensions. The initial set of parameter estimates would be expanded to three (*i.e.* $n+1$), and the centroid would consist of a straight line connecting the two best points. It will then try estimates on a vector from the worst point orthogonal to the straight line in the direction that gives lower values of the least-squares function. When this direction no longer produces optimization, it will use the new set of three values for another round of iteration.

3.5. Confidence Limits

After the optimization of the parameter values is concluded, most programs will estimate the standard deviation of each. This is a particularly valuable feature since it allows some rational basis for the comparison of the values under different conditions. Unfortunately there are several methods for calculation of these estimates, each of which involves somewhat different assumptions such as the extent to which the parameters are correlated with each other and whether the dependent variable is a linear function of the parameter in question in the vicinity of the minimum. It is beyond the scope of this book to discuss the various methods and the interested reader is referred to the article by Johnson and Faunt [1] as well references cited therein.

3.6. Model Comparison

There are several intuitive methods for model comparison and a few systematic ones. Generally in steady-state enzyme kinetics the model equations are all within a family of models that are related to each other by the presence or absence of an additional parameter. Unnecessary parameters in a model frequently go to zero (or sometimes negative) as a result of the optimization. Parameters may also have large standard errors as a result of the presence of unnecessary parameters or a suboptimum model.

In addition the best model generally has the lowest least-squares function value at the minimum. However, this comparison is complicated somewhat by differences in the number of degrees of freedom for each model, *i.e* the number of data points minus the number of parameters optimized. Some of the available programs calculate a statistical quantity that includes compensation for the number of degrees of freedom and can be used to compare models. However, there is a residual conviction among investigators in the field that if you have to resort to this, you probably will not convince many other people.

3.7. Utilization of Available Software

Although it is possible to write your own program for nonlinear curve fitting a number of good commercial programs are available. Some of the software requires the input of the analytical differentials with respect to each of the parameters to be optimized but contemporary programs accomplish this numerically. Also some will accept only a limited population of mathematical models for steady-state enzyme kinetics, whereas other programs are more general and will accept any equation the operator can write.

Some searching on the internet will reveal a large number of satisfactory programs, some of which are free. Currently there are three types of curvefitting programs available: general curvefitting programs that are part of a larger statistical package; general curve fitting programs that are stand alone programs, but usually include some graphics; and specific curvefitting programs for enzyme kinetics.

The large statistical packages such as Matlab® and MLAB® (Table 3.1) are very versatile but expensive and usually require the operator to write some program steps to call the routines.

The second category, the stand-alone general curvefitting programs such as Scientist®, Fitall®, and CurveExpert 1.3® (Table 3.1), contain no statistical or algebraic routines unrelated to their primary purpose. They will fit data to any equation you want to write but require the operator to write enough program steps to specify the model equation, the variables, the parameters, and their initial values. Some of these programs will do numeric integration of differential equations, referred to in a later chapter.

The third category, the specific programs for enzyme kinetics, are the easiest to operate but some include equations for only the most common kinetic models, whereas other programs allow the operator to write some additional mathematical models with which to search.. Some of these programs will find preliminary estimates for initial parameter values.

TABLE 3.1. Available Software for Data Fitting.

Program	Source	Remarks
Scientist	http://www.micromath.com	General curve-fitting
FitAll	htt://home.ican.net/~mtrsoft/abtfa w.htm	General curve-fitting
Dynafit	http://www.biokin.com	Enzyme fitting, PC & Mac, Free
EnzFitter	http://www.biosoft.com	Enzyme fitting
EZ-Fit	http://www.jlc.net/~fperrell/webpx 05.htm	Enzyme fitting
SigmaPlot 2000	http://www.spssscience.com	Graphics package, Enzyme module
MLAB	http://www.civilized.com	Math/statistical package
Matlab	http://www.math.utah.edu/lab/ms/ matlab/matlab.html	Math/statistical package
CurveExpert 1.3	http://www.ebicom.net/~dhyams/c vxpt.htm	General curve-fitting

3.8. The Purpose of Mathematical Models in Different Forms

The actual data fitting and model evaluation is done with the mathematical model for initial velocity, v_i, as the dependent variable and the concentration of substrate, inhibitor and product, if any, as the independent variable. The independent variables should not appear in a term such that the term will be infinite, if the concentration equals zero, since most programs will fail under these circumstances. For example the simple Michaelis-Menten equation would be fit as equation 3.7.

$$v_i = \frac{V_{max}*(A)}{K_M+(A)}$$

(3.7)

Much of the data and model descriptions in the remainder of the book will be in terms of the slopes and intercepts of the double-reciprocal plot (*e.g.* equation 3.8).

$$\frac{1}{v_i} = \frac{K_M}{V_{max}}*\frac{1}{(A)}+\frac{1}{V_{max}}$$

(3.8)

Because of the convenience of derivation and as a compromise most of the mathematical models in the remainder of the book will be expressed with v_i as the dependent variable and the right side of the equation as the reciprocal (*e.g.* equation 3.9).

$$v_i = \frac{1}{\dfrac{K_M}{V_{max}}*\dfrac{1}{(A)}+\dfrac{1}{V_{max}}}$$

(3.9)

3.9. Summary

Mathematical modeling is used to determine which chemical model agrees best with the data and to estimate the values of the kinetic parameters. Although it is useful to use graphical methods for preliminary data display and model testing, reiterative methods by computer provide a more rigorous approach and provide estimates of the error associated with each kinetic parameter. The most common reiterative methods minimize the value of the least-squares function. The available computer programs utilize a combination of the methods.

3.10. References

1. Johnson, M.L. and Faunt, L.M. "Parameter Estimation by Least-Squares Methods," *Methods Enzymol.* 210, 1-37 (1992).

2. Straume, M. and Johnson, M.L. "Analysis of Residuals: Criteria for Determining Goodness-of-Fit," *Methods Enzymol.* 210, 87-105.

3. Wilkinson, G.N. "Statistical Estimations in Enzyme Kinetics," *Biochem. J.* 80, 324-32 (1961).

4. Beechem, J.M. "Global Analysis of Biochemical and Biophysical Data," *Methods Enzymol.* 210, 38 (1992).

5. Cleland, W.W., "Statistical Analysis of Enzyme Kinetic Data," *Methods Enzymol.* 63, 103-138 (1979).

3.11. Appendix 3.1: Derivation of Least-Squares, Polynomial-Fitting Algorithm

The objective is to fit the equation 3.10 to data set x_i, y_i.

$$y = a * x^2 + b * x + c \tag{3.10}$$

The least-squares function is:

$$\sum (r^2) = \sum (y_i - a * x_i^2 - b * x_i - c)^2 \tag{3.11}$$

The first derivative of the least-squares function with respect to a is:

$$\frac{d(\sum (r^2))}{da} = 2 * \sum (y_i - a * x_i^2 - b * x_i - c) * \frac{d}{da} * (y_i - a * x_i^2 - b * x_i - c)$$

$$= 2 * \sum (y_i - a * x_i^2 - b * x_i - c) * (-x_i^2) \tag{3.12}$$

Set the previous equation equal to zero.

$$0 = \sum (-y_i * x_i^2 + a * x_i^4 + b * x_i^3 + c * x_i^2)$$

$$0 = -\sum y_i * x_i^2 + a * \sum x_i^4 + b * \sum x_i^3 + c * \sum x_i^2 \tag{3.13}$$

Repetition of the same process for the parameter b.

$$0 = -\sum y_i * x_i + a * \sum x_i^3 + b * \sum x_i^2 + c * \sum x_i \tag{3.14}$$

Repetition of the same process for the parameter c.

$$0 = -\sum y_i + a * \sum x_i^2 + b * \sum x_i + n * c \tag{3.15}$$

The last three equations constitute a series of linear simultaneous equations that can be solved for a, b, and c.

CHAPTER 4

DERIVATION OF MATHEMATICAL MODELS

4.1. Introduction

This chapter contains a discussion of the methods by which mathematical models for initial velocity can be derived from the chemical models for the mechanism of the reaction. The derivation of the Michaelis-Menten equation for a simple, single-substrate enzymatic reaction that appears in most textbooks on biochemistry can be followed rather easily, but more complicated chemical models require a more systematic approach. Unfortunately most of the systematic methods have a very precise structure and are somewhat complicated themselves.

Although there have been several attempts to interpret steady-state enzyme kinetic data and graphic plots of data in an intuitive manner without reference to mathematical models, it has been generally unconvincing and requires more time and energy than their derivation. In contrast other presentations of the subject have contained derivations of mathematic models for every possible chemical model.

In the present chapter the same mathematical model will be derived for the simple, single-substrate chemical model by three different methods. One of the methods requires some effort to understand its logic, but is very easy to use. With it the mathematical models can be written by inspection of the chemical model. A more detailed discussion of additional methods for the derivation of mathematical models can be found in the chapter by Huang [1].

4.2. The Conventions of Notation

The discussion of the derivation methods will be facilitated by an explanation of the context and the notations that will be used.

$$E \underset{k_2}{\overset{k_1}{\rightleftharpoons}} EA \underset{k_4}{\overset{k_3}{\rightleftharpoons}} EP \overset{k_5}{\longrightarrow} E$$

Figure 4.1

The chemical model for an enzyme-catalyzed reaction that is ordinarily presented in the context of a linear process (e.g. Figure 4.1) will be considered as a catalytic cycle and written as closed polygonal figures with the various enzyme intermediates at the vertices. For example the same chemical model in Figure 4.1 will be considered in the context of Figure 4.2. However, later there may be various appendages from one or more vertices denoting dead-end intermediates. The forward direction of the cycle is defined by the

27

irreversible step.   Therefore, the downstream and
upstream directions of the cycle are defined.

The rate constants, k, for the downstream
steps, *i.e.* forward steps,   have odd numbers as
subscripts and the upstream rate constants have even
numbers. The equilibrium constants for the reaction
steps, K, have odd numbers as subscripts corresponding
to the number of the forward rate constant for that step.

The initial velocity is v_i. The maximum
velocity, V_{max}, is the initial velocity when all substrates
are at infinite concentration. The units of velocity are

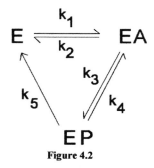

Figure 4.2

generally concentration change per time, *e.g.*
molar/min. The precise mechanistic definition depends upon the chemical model for the
reaction. It's mechanistic meaning will be discussed in more detail later. In a number of
contexts it is easier to calculate and discuss the value of k_{cat}, which is the V_{max} divided by
the enzyme concentration in the same concentration units. The unit for K_{cat} is inverse time,
e.g. sec⁻¹.

The substrates are denoted by A, B, C, etc. The latter are in parentheses to denote
concentration. The Michaelis constant in a global context is K_M, but in a context with a
specific substrate is K_A, K_B, K_C, etc. The precise mechanistic definition depends upon the
chemical mechanism and will be discussed later. The operational definition is that the K_M
is the substrate concentration that gives half the maximum velocity, when all other
substrates, if any, are at infinite concentration. Product is denoted by P, Q, R, etc. The latter
are in parentheses to denote concentration.

4.3. Methods for Derivation

Three methods for the derivation of mathematical models will be discussed: derivation by
determinants, the King-Altman method and derivation by inspection. Each method will
be demonstrated by derivation of the mathematical model for the same chemical Model.

The chemical model is one with a single substrate, an enzyme-substrate complex
and an enzyme-product complex (Figure 4.1). Since enzyme-catalyzed reactions are
essentially cyclic, from the point of view of the enzyme, the same mechanism is written as
a closed polygon, and the substrate concentration will be expressed with the rate constants
(Figure 4.2). However, in the interest of simplicity the substrate term will be omitted for
the present. It will be necessary to remember that every term in the final mathematical
model that contains the rate constant k_1 must be multiplied by the concentration of
substrate.

4.3.1. DERIVATION BY DETERMINANTS

The most rigorous method for derivation is to solve a series of simultaneous linear equations by determinants or by repetitive and judicial substitution of one equation into another. Although it is also probably the most laborious, it does demonstrate the critical assumptions and approximations in steady-state enzyme kinetics. In order to start, write the equation for the rate of change in concentration of each of the enzyme intermediates but one; an equation for the rate of product formation: and the equation for the conservation of enzyme, equations 4.1-4.4 respectively.

$$\frac{d(EA)}{dt} = k_1*(E) + k_4*(EP) - (k_3 + k_2)*(EA) \tag{4.1}$$

$$\frac{d(EP)}{dt} = k_3*(EA) - (k_4 + k_5)*(EP) \tag{4.2}$$

$$\frac{d(P)}{dt} = v_i = k_5*(EP) \tag{4.3}$$

$$(E_t) = (E) + (EA) + (EP) \tag{4.4}$$

The enzyme conservation equation provides the logical relationship for the rate of change of the enzyme intermediate for which no explicit equation was written.

Since the principal assumption of steady-state enzyme kinetics is that the concentration of the enzyme intermediates is constant during the initial velocity measurement, the equations for the rate of change of these intermediates can be set equal to zero (equations 4.5). Furthermore, the equation for the rate of product formation can be rearranged. These manipulations result in a set of four simultaneous equations in four unknowns, (E), (ES), (EP) and v_i (equation 4.5).

$$
\begin{array}{ccccl}
k_1*(E) & -(k_2+k_3)*(ES) & +k_4*(EP) & +0 & =0 \\
0 & +k_3*(ES) & -(k_4+k_5)*(EP) & +0 & =0 \\
0 & +0 & -k_5*(EP) & +v_i & =0 \\
(E) & +(ES) & +(EP) & +0 & =E_t
\end{array}
\tag{4.5}
$$

This set of equations can be solved for v_i by any one of several methods. Here the determinant ratio will be solved by Cramer's rule. Simultaneous equations can be solved for a given unknown by a ratio of two matrices (determinants) the denominator of which is the coefficients of the unknown terms in the simultaneous equations and the numerator matrix is the same as the denominator except that the column containing the coefficients of the unknown term, *i.e.* v_i in our example, for which the set is being solved is substituted by the column of terms that contain no unknown terms (the column on the right of the equal sign in our example), equation 4.6. Each of the matrices can be evaluated by

$$v_i = \frac{\begin{vmatrix} k_1 & -(k_2+k_3) & +k_4 & 0 \\ 0 & +k_3 & -(k_4+k_5) & 0 \\ 0 & 0 & -k_5 & 0 \\ 1 & +1 & +1 & E_i \end{vmatrix}}{\begin{vmatrix} k_1 & -(k_2+k_3) & +k_4 & 0 \\ 0 & +k_3 & -(k_4+k_5) & 0 \\ 0 & 0 & -k_5 & 1 \\ 1 & +1 & +1 & 0 \end{vmatrix}} \tag{4.6}$$

standard methods described in texts on algebra and in section 4.6, Appendix 4.1 (equations 4.7-4.10), and result in equation 4.11. Before the meaning of the equation is discussed the

$$v_i = -E_i * \frac{\begin{vmatrix} k_1 & -(k_2+k_3) & +k_4 \\ 0 & +k_3 & -(k_4+k_5) \\ 0 & 0 & -k_5 \end{vmatrix}}{\begin{vmatrix} k_1 & -(k_2+k_3) & +k_4 \\ 0 & +k_3 & -(k_4+k_5) \\ 1 & +1 & +1 \end{vmatrix}} \tag{4.7}$$

same result will be produced by the other two methods.

$$v_i = -E_t * k_1 * \frac{\begin{vmatrix} k_3 & -(k_4+k_5) \\ 0 & -k_5 \end{vmatrix}}{k_1 * \begin{vmatrix} k_3 & -(k_4+k_5) \\ 1 & +1 \end{vmatrix} + \begin{vmatrix} -(k_2+k_3) & +k_4 \\ k_3 & -(k_4+k_5) \end{vmatrix}}$$

(4.8)

$$v_i = \frac{E_t * k_1 * k_3 * k_5}{k_1 * k_3 + k_1 * (k_4+k_5) + (k_2+k_3)(k_4+k_5) - k_3 * k_4}$$

(4.9)

$$v_i = \frac{k_5 * E_t}{1 + \dfrac{k_4+k_5}{k_3} + \dfrac{k_2 * k_4 + k_2 * k_5 + k_3 * k_5}{k_1 * k_3}}$$

(4.10)

$$\frac{v_i}{E_t} = \frac{1}{\dfrac{1}{k_5} + \dfrac{k_4}{k_5 * k_3} + \dfrac{1}{k_3} + \dfrac{k_2 * k_4}{k_1 * k_3 * k_5} + \dfrac{k_2}{k_1 * k_3} + \dfrac{1}{k_1}}$$

(4.11)

4.3.2. THE KING-ALTMAN METHOD

The King-Altman method is named for the authors who first described it [2] and is operationally a graphical method, although it is based on the determinant method. However, this method has been used extensively by enzyme kineticists for decades, and is probably the standard by which other methods are validated. In order to implement this method start with the polygonal form of the chemical mechanism (Figure 4.2). Write all of the subforms of this figure in which all of the enzyme forms are connected but which contain no closed polygons (Figure 4.3).

From these one can write an equation for the ratio of the concentration of each

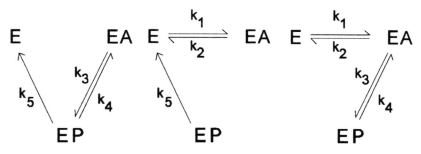

Figure 4.3

enzyme form to the concentration of the total enzyme. The numerator of the ratio is the sum of each of the products of all of the rate constants of steps leading toward the enzyme form under consideration, one product for each graphical subform of the original chemical model. Thus, for this model there should be three subforms. The denominator is the sum of all of the similar terms for all of the enzyme intermediates.

For example the fraction of the total enzyme that is complex EP is given in equation 4.12, where the denominator is indicated by a summation sign, $\sum$. It will be written later. The numerator terms for the first and second subforms are zero because the step between E and EP is irreversible and the rate constant from E to EP is zero. The equations for the remaining enzyme intermediates follow the same rules (equations 4.13 and 4.14) for the numerators. The denominator for each of the three equations is the sum of the numerators of all three (equation 4.15).

$$\frac{(EP)}{(E_t)} = \frac{0+0+k_1*k_3}{\sum} \tag{4.12}$$

$$\frac{(EA)}{(E_t)} = \frac{k_1*k_5+k_1*k_4+0}{\sum} \tag{4.13}$$

$$\frac{(E)}{(E_t)} = \frac{k_2*k_5+k_2*k_4+k_3*k_5}{\sum} \tag{4.14}$$

$$\sum = k_1*k_3+k_1*k_5+k_1*k_4+k_2*k_5+k_2*k_4+k_3*k_5 \tag{4.15}$$

The initial velocity of the enzyme-catalyzed reaction is the rate constant of the irreversible step multiplied by the concentration of the immediately upstream enzyme form. For example the initial velocity for the chemical model under consideration is given in equation 4.16.

$$v_i = k_5 * (EP) \tag{4.16}$$

Substitution of the equation (equation 4.12) for the fraction of total enzyme that is the proximal intermediate of the irreversible step, EP in our example, into the velocity equation (equation 4.16) yields equation 4.17,

$$\frac{v_i}{(E_t)} = k_5 * \frac{k_1 * k_3}{k_1 * k_3 + k_1 * k_5 + k_1 * k_4 + k_2 * k_5 + k_2 * k_4 + k_3 * k_5} \tag{4.17}$$

which can be rearranged to equation 4.18.

$$\frac{v_i}{E_t} = \frac{1}{\dfrac{1}{k_5} + \dfrac{k_4}{k_5 * k_3} + \dfrac{1}{k_3} + \dfrac{k_2 * k_4}{k_1 * k_3 * k_5} + \dfrac{k_2}{k_1 * k_3} + \dfrac{1}{k_1}} \tag{4.18}$$

Comparison of equation 4.18 with equation 4.11 reveals that they are identical. This method becomes rather complicated and laborious with more complicated chemical models and there are formulas for the calculation of the number of subforms of the original geometric model and the number of possible terms in the numerator of the equations. The reader is referred to more detailed references for these and other more detailed information [1][3].

4.3.3. DERIVATION BY INSPECTION

Derivation by inspection is the method that will be used in the remainder of this book, because it is the most rapid and the most easily implemented. Although it can be described rather briefly, it is more convincing is we start with Clelend's "net-rate-constant" method [4]. Start with the same chemical model in Figure 4.2. However, in the net-rat-constant method it is recognized that since a proportion of the downstream flux in each step actually goes on to product, an irreversible rate constant is associated with each step. Assume initially that all of the steps are irreversible as in Figure 4.4 and, represent the rate constants as the net rate constants, k'. The reversibility of two of them will be acknowledged later.

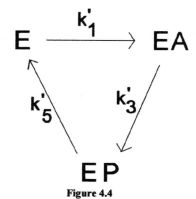

Figure 4.4

The steady-state approximation specifies that the rates of all three steps are equal to each other. That rate is also equal to the initial velocity (equation 4.19). Division of each of the expressions in equation 4.19 by (E_t) results in the set of equations 4.20.

Division of both sides of each equation

$$v_i = k_1' * (E) = k_3' * (EA) = k_5' * (EP) \tag{4.19}$$

$$\frac{v_i}{E_t} = \frac{k_1' * (E)}{E_t}$$

$$\frac{v_i}{E_t} = \frac{k_3' * (EA)}{E_t} \tag{4.20}$$

$$\frac{v_i}{E_t} = \frac{k_5' * (EP)}{E_t}$$

in the set 4.20 by the net rate constant results in the equations 4.21.

$$\frac{1}{k_1'} * \frac{v_i}{E_t} = \frac{(E)}{E_t}$$

$$\frac{1}{k_3'} * \frac{v_i}{E_t} = \frac{(EA)}{E_t} \tag{4.21}$$

$$\frac{1}{k_5'} * \frac{v_i}{E_t} = \frac{(EP)}{E_t}$$

The sum of the three equations 4.21 is expressed in 4.22. Because of the

$$\frac{v_i}{E_t}*(\frac{1}{k_1'}+\frac{1}{k_3'}+\frac{1}{k_5'})=\frac{((E)+(EA)+(EB))}{E_t} \tag{4.22}$$

conservation of enzyme the sum in equation 4.22 is equal to 1.0 (equation 4.23).

$$\frac{v_i}{E_t}*(\frac{1}{k_1'}+\frac{1}{k_3'}+\frac{1}{k_5'})=1.0 \tag{4.23}$$

Finally division of both sides of equation 4.23 by the sum of the reciprocals of the net rate constants results in equation 4.24.

$$\frac{v_i}{E_t}=\frac{1}{\dfrac{1}{k_1'}+\dfrac{1}{k_3'}+\dfrac{1}{k_5'}} \tag{4.24}$$

Equation 4.24 expresses the initial velocity as the reciprocal of the sum of the reciprocals of the net rate constants.

Now it is necessary to examine the implications of reversibility. Each of the irreversible rate constants is actually the rate constant of the flux through that step that actually completes the cycle or goes on to product. The relationship of the irreversible and the actual rate constant k_5 is simple, because the step is already irreversible. Therefore:

$$k_5'=k_5 \tag{4.25}$$

The irreversible flux through the step represented by k_3 is the total flux from EA to EP times the ratio of that flux that actually goes on through the irreversible step with the rate constant k_5 (equation 4.26). Thus the ratio is the amount of flux through the step with the rate constant k_5 divided by the sum of that flux plus the reverse flux back to EA. In the steady state the sum in the denominator is equal to the total flux from EA to EP, both the forward and the reverse flux from EP.

$$k_3' * (EA) = k_3 * (EA) * \left[\frac{k_5 * (EP)}{k_5 * (EP) + k_4 * (EP)} \right] \qquad \text{(4.26)}$$

When the enzyme terms (EA) and (EP) are eliminated from equation 4.26, the result is equation 4.27.

$$k_3' = k_3 * \left[\frac{k_5}{k_5 + k_4} \right] \qquad \text{(4.27)}$$

Take the reciprocal of k_3' to arrive at equation 4.28.

$$\frac{1}{k_3'} = \frac{1}{k_3} + \frac{k_4}{k_3 * k_5} \qquad \text{(4.28)}$$

This equation will be revisited, when we finish a similar operation for k_1'. The rate constant k_1' is that portion of the total forward flux through the step represented by the rate constant k_1 that actually completes the reaction. Therefore, it will be the total forward flux times the fraction that goes on to the end, equation 4.29.

$$k_1' * (E) = k_1 * (E) * \left[\frac{k_3' * (EA)}{k_3' * (EA) + k_2 * (EA)} \right] \qquad \text{(4.29)}$$

Elimination of the enzyme terms results in equation 4.30.

$$k_1' = k_1 * \left[\frac{k_3'}{k_3' + k_2} \right] \qquad \text{(4.30)}$$

Substitution of equation 4.27 above for k_3' into equation 4.30 results in equation 4.31.

$$k_1' = k_1 * \left[\frac{k_3}{\left[k_3 * \left[\dfrac{k_5}{k_5+k_4} \right] + k_2 \right]} \right] * \left[\frac{k_5}{k_5+k_4} \right]$$

$$k_1' = \left[\frac{k_1 * k_3 * k_5}{\left[\dfrac{k_3 * k_5}{k_5+k_4} + k_2 \right] * [k_5+k_4]} \right]$$

(4.31)

Rearrangement results in equation 4.32.

$$k_1' = \frac{1}{\dfrac{1}{k_1} + \dfrac{k_2 * k_4}{k_1 * k_3 * k_5} + \dfrac{k_2}{k_1 * k_3}}$$

(4.32)

The reciprocal of equation 4.32 is equation 4.33

$$\frac{1}{k_1'} = \frac{1}{k_1} + \frac{k_2 * k_4}{k_1 * k_3 * k_5} + \frac{k_2}{k_1 * k_3}$$

(4.33)

.
Finally substitution of equations 4.25, 4.28 and 4.33 into equation 4.24 results in equation 4.34 for initial velocity in terms of the individual rate constants. A little inspection of equation 4.34 reveals that it is identical to equation 4.11 and equation 4.18.

$$\frac{v_i}{E_t} = \frac{1}{\dfrac{1}{k_5} + \dfrac{k_4}{k_5 * k_3} + \dfrac{1}{k_3} + \dfrac{k_2 * k_4}{k_1 * k_3 * k_5} + \dfrac{k_2}{k_1 * k_3} + \dfrac{1}{k_1}}$$

(4.34)

Since an algorithm for derivation by inspection is not yet apparent, some additional steps are necessary. Remember that each of the steps in our chemical model also has an equilibrium constant expressed by K_1 and K_3 (capital K instead of small k) below (equation 4.35).

$$K_1 = \frac{k_1}{k_2}$$

$$K_3 = \frac{k_3}{k_4}$$

(4.35)

Substitution of these into equation 4.28 and equation 4.33 for the appropriate ratios of rate constants results in equation 4.36.

$$\frac{1}{k_3'} = \frac{1}{k_3} + \frac{1}{K_3 * k_5}$$

$$\frac{1}{k_1'} = \frac{1}{k_1} + \frac{1}{K_1 * K_3 * k_5} + \frac{1}{K_1 * k_3}$$

(4.36)

Substitution of these into equation 4.11, 4.18 or 4.34 for the appropriate ratios of rate constants and a little rearranging of the denominator terms plus insertion of the terms for substrate concentration results in equation 4.37.

$$\frac{v_i}{E_t} = \frac{1}{\dfrac{1}{k_1} + \dfrac{1}{K_1 * k_3} + \dfrac{1}{K_1 * K_3 * k_5} + \dfrac{1}{k_3} + \dfrac{1}{K_3 * k_5} + \dfrac{1}{k_5}}$$

$$\frac{v_i}{E_t} = \frac{1}{\dfrac{1}{(A)} * \left[\dfrac{1}{k_1} + \dfrac{1}{K_1 * k_3} + \dfrac{1}{K_1 * K_3 * k_5} \right] + \dfrac{1}{k_3} + \dfrac{1}{K_3 * k_5} + \dfrac{1}{k_5}}$$

(4.37)

Derivation by inspection of the expression for $v_i/(E_t)$ is to write the reciprocal of the sum of the reciprocals of each forward rate constant plus a sum of terms composed of the reciprocal of one forward rate constant at a time times the adjacent upstream equilibrium constants, one additional equilibrium constant in each term, until the next irreversible step (reciprocal equilibrium constant of the irreversible step is zero).

In the context of a polygonal chemical model of n steps, m of which are irreversible the equation for $v_i/(E_t)$ is the reciprocal of a sum of sums, sum a and sum b respectively. Each term is a reciprocal product of a rate constant multiplied by zero to n-m equilibrium constants of adjacent upstream steps. Sum b is a series of the reciprocal products each of which contains the same rate constant times zero to n-m equilibrium constants of adjacent upstream steps. Each subsequent term in the series contains the next additional upstream equilibrium constant than the previous term. If one of the upstream steps is irreversible the reciprocal of the equilibrium constant is zero and the series is terminated. Sum a is the sum of sum b elements each of which contains the reciprocal of a different forward rate constant.

Although it does not seem particularly helpful in derivation, equation 4.38 is an attempt at a formal statement of the algorithm for a chemical model with "n" steps written in polygonal format in which the step represented by the rate constant with the highest number is irreversible. The index "i" refers only to forward rate constants.

$$\frac{v_i}{E_t} = \frac{1}{\sum_{i=1}^{i=n}\left[\frac{1}{k_i} + \sum_{j=i-1}^{j=1}\left[\prod_{i-1}^{j}\frac{1}{K_j}\right]*\frac{1}{k_i}\right]} \tag{4.38}$$

Somewhat more useful is an operational description as a flow diagram (Figure 4.5). First write the left side of the equation, $v_i/(E_t)$. Second go to the right side. Write 1 with a line under it. Third, write the reciprocal of the rate constant of one of the irreversible steps. Fourth, add to the denominator another copy of the same term but multiplied by the reciprocal of the equilibrium constant of the step immediately upstream (unless it is equal to zero). Fifth, repeat the third step unless the last term was zero (the reciprocal of the last additional equilibrium constant was zero). Sixth, carry out the second, third and fourth processes with any forward rate constant that have not yet been used. In words the foregoing sounds like a rather long and complicated derivation. However, with very little practice it goes about as fast as one can write the terms.

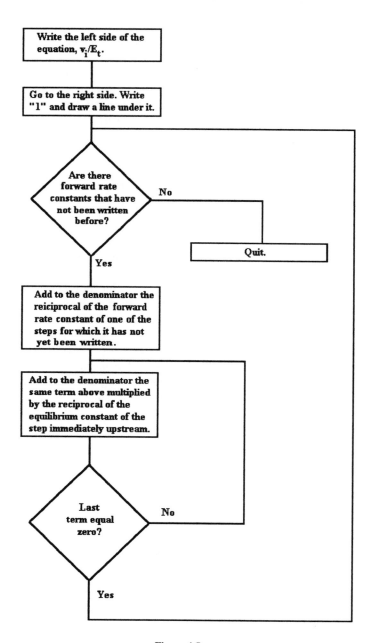

Figure 4.5

For practice it might be useful to write the mathematical model for a chemical model with four enzyme intermediates instead of three (Figure 4.6). The mathematical model should look like equation 4.39. Moreover the logic for the model is similar to that presented above for the three-step model.

It is now appropriate to regard the concentration of substrate, A, in the first chemical model. Since substrate was previously included with the rate constant k_1 in the chemical model, it will now be multiplied times every term in the mathematical model that contains k_1. Of course, since the terms that contain K_1 also contain k_1, the former

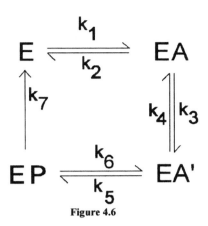

Figure 4.6

should also be multiplied times the substrate concentration. The equation for the chemical

$$\frac{v_i}{E_t} = \frac{1}{\dfrac{1}{k_7} + \dfrac{1}{K_5 * k_7} + \dfrac{1}{K_3 * K_5 * k_7} + \dfrac{1}{K_1 * K_3 * K_5 * k_7} + \dfrac{1}{k_5} + \dfrac{1}{K_3 * k_5} + \dfrac{1}{k_3} + \dfrac{1}{K_1 * K_3 * k_5} + \dfrac{1}{K_1 * k_3} + \dfrac{1}{k_1}} \qquad (4.39)$$

model (Figure 4.1) relating the initial velocity to substrate concentration should look like equation 4.40.

$$\frac{v_i}{E_t} = \frac{1}{\dfrac{1}{k_5} + \dfrac{1}{K_3 * k_5} + \dfrac{1}{k_3} + \left[\dfrac{1}{K_1 * K_3 * k_5} + \dfrac{1}{K_1 * k_3} + \dfrac{1}{k_1}\right] * \dfrac{1}{(A)}} \qquad (4.40)$$

The operational and mechanistic significance of the various assemblages of rate and equilibrium constants will be discussed in the following chapter.

4.4. Summary

Three methods for the derivation of kinetic mathematical models from the chemical models have been described: by determinants, by the King-Altman method and by inspection.

Because of its simplicity and ease of implementation the third method will be used exclusively in the remainder of this book. This method has the limitation that it does not handle branching in any convenient way, similar to the limitations of the related method of net-rate-constants described by Cleland [4]. This limitation will first become apparent in the development of the mathematical model for a bireactant, steady-state, random chemical model. However, this limitation does not hinder its use for the derivation of mathematical models for most of the chemical models commonly encountered.

4.5 References

1. Huang, C.Y., "Derivation of Initial Velocity and Isotope Exchange Rate Equations," *Methods Enzymol.* 63, 54-84 (1979).

2. King, E.L. and Altman, C. "A Schematic Method of Deriving the Rate Laws for Enzyme-Catalyzed Reactions," *J. Phys. Chem*, 60, 1375-8 (1956).

3.Segal, I.H., *Enzyme Kinetics*, Wiley Interscience, New York, p. 506-523 (1975)

4. Cleland, W.W., "Partition Analysis and the Concept of Net Rate Constants as Tools in Enzyme Kinetics," *Biochemistry*, 14, 3220-4 (1975).

4.6. Appendix 4.1: Evaluation of Matrices

A determinant of order n can be reduced to several of order n-1 by multiplication of each of the elements of a row or column by the determinant made up of the elements that are excluded from the row and column of that particular element. The upper left element and every-other one in its column and row is positive. The remainder in its column and row are negative. For example:

$$\begin{vmatrix} a_{11} & a_{12} & a_{13} & a_{14} \\ a_{21} & a_{22} & a_{23} & a_{24} \\ a_{31} & a_{32} & a_{33} & a_{34} \\ a_{41} & a_{42} & a_{43} & a_{44} \end{vmatrix} = a_{12}*\begin{vmatrix} a_{22} & a_{23} & a_{24} \\ a_{32} & a_{33} & a_{34} \\ a_{42} & a_{43} & a_{44} \end{vmatrix} - a_{21}*\begin{vmatrix} a_{12} & a_{13} & a_{14} \\ a_{32} & a_{33} & a_{34} \\ a_{42} & a_{43} & a_{44} \end{vmatrix} +$$

$$a_{31}*\begin{vmatrix} a_{12} & a_{13} & a_{14} \\ a_{22} & a_{23} & a_{24} \\ a_{42} & a_{43} & a_{44} \end{vmatrix} - a_{41}*\begin{vmatrix} a_{12} & a_{13} & a_{14} \\ a_{22} & a_{23} & a_{24} \\ a_{32} & a_{33} & a_{34} \end{vmatrix}$$

(4.41)

By application of the same procedure:

$$\begin{vmatrix} a_{22} & a_{23} & a_{24} \\ a_{32} & a_{33} & a_{34} \\ a_{42} & a_{43} & a_{44} \end{vmatrix} = a_{22}*\begin{vmatrix} a_{33} & a_{34} \\ a_{43} & a_{44} \end{vmatrix} - a_{32}*\begin{vmatrix} a_{23} & a_{24} \\ a_{43} & a_{44} \end{vmatrix} + a_{42}*\begin{vmatrix} a_{23} & a_{24} \\ a_{33} & a_{34} \end{vmatrix}$$

(4.42)

Furthermore:

$$\begin{vmatrix} a_{33} & a_{34} \\ a_{43} & a_{44} \end{vmatrix} = a_{33}*a_{44} - a_{43}*a_{34}$$

(4.43)

By repetitive application of these rules the initial matrix above can be evaluated.

CHAPTER 5

THE EFFECT OF SUBSTRATE CONCENTRATION

5.1. Introduction

Different enzymes catalyze reactions in which different numbers of substrates react with each other to yield different numbers of products. A number of different possible chemical models will describe multiple substrates reacting with each other and the release of multiple products. Data from several kinds of experiments will permit these chemical models to be distinguished from each other. The objectives of the present chapter are to describe the possible results of experiments in which initial velocity is measured in the presence of various concentrations of the substrates as well as to describe the logical process relating these results to the possible chemical models for the reaction. Generally only selected models for the various sequences of multiple substrate binding can be distinguished with the results of these experiments.

5.2. Enzyme Classification

In order to gain a global perspective before individual models are presented it is useful to discuss enzyme classification. Enzymes are formally classified into six classes (Table 1.1) according to the chemistry of the overall reaction catalyzed.

Table 5.1. The Classification of Enzymes.

Name	Description of Catalysis
1. oxidoreductase	Electrons transferred from one substrate to another.
2. transferase	Chemical group transferred from one substrate to another.
3. hydrolase	Water is added across a chemical bond to break it.
4. lyase	A double bond is made or destroyed.
5. isomerase	Atoms are rearranged.
6. ligase	Molecules are attached to each other with an energy expense.

In order to distinguish chemical kinetic models Cleland [1] offered a classification according to the number of substrates and products that participate in the overall reaction. For example reactions are classified as uni:uni, uni:bi, bi:bi, or ter:bi, if they have a single substrate and product, a single substrate and two products two substrates and two products or three substrates and two products. In the directions indicated the first two examples would be catalyzed by a unireactant enzymes. The third would be a by a bireactant enzyme, and the fourth would be catalyzed by a terreactant enzyme.

There is a general correlation between the formal class of enzyme and the number of substrate and product species, but, of course there are individual exceptions, particularly with enzymes whose formal classification is somewhat ambiguous. Isomerases, for example, have one substrate and one product. In the nomenclature of Cleland [1] they are called uni:uni reactions. Hydrolases as a class of enzymes kinetically also have one substrate but two products, since the concentration of water does not vary in most experiments. The latter are called uni:bi reactions.

Most of the lyase class of enzymes are uni:bi or bi:uni depending upon the direction in which they are regarded. Alternatively they may be kinetically uni:uni, if water is added to a double bond, *e.g.* fumarase. Most of the transferase class of enzymes, of which the kinases are a subclass are bi:bi reactions as are most of the simple dehydrogenases, a subclass of the oxidoreductases,. Finally most of the ligase class of enzymes catalyze ter:ter reactions because two molecules form an adduct and a high energy compound, usually ATP, is cleaved, usuallly to ADP and inorganic phosphate. Thus the enzyme has three substrates and three products.

5.3. Chemical Models and Mathematical Models

In this section mathematical models will be derived for the initial velocity of the most common chemical models and those that can be distinguished from each other will be pointed out. Although the actual fitting of experimental data would be carried out by computer as described in Chapter 3, the discussion below will be in the context of double reciprocal plots, because it is more convenient conceptually to discuss differences in slopes and intercepts. The mathematical models are presented in the format below (*e.g.* first equation 1.1), in an effort to make these relationships more apparent. The actual fitting to data is done with the model in a somewhat different format, one with operational parameters (K_M, V_m, V_m/K_M; *e.g.* second equation 1.1) and with substrate concentration in the numerator, since the terms containing reciprocal substrate concentration frequently cause the programs to abort at zero substrate concentration.

5.3.1. UNIREACTANT ENZYMES

Models of Single-substrate reactions are generally less interesting than those of multiple substrate reactions because the only ones of the former that can be written reasonably cannot be distinguished from each other in experiments in which only the concentration of substrate is varied. However, a discussion of them helps provide some understanding of the meaning of the kinetic parameters, V_{max}, k_{cat}, V_{max}/K_M, k_{cat}/K_M, and K_M, with which we are more familiar.

A chemical model for a reaction with a single substrate (Figure 5.1) was presented and the corresponding mathematical model (equation 5.1) was derived in the previous chapter.

The maximum velocity, V_{max}, of an

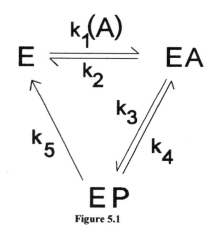

Figure 5.1

$$\frac{v_i}{E_t} = \frac{1}{\dfrac{1}{k_5} + \dfrac{1}{K_3 * k_5} + \dfrac{1}{k_3} + \left[\dfrac{1}{K_1 * K_3 * k_5} + \dfrac{1}{K_1 * k_3} + \dfrac{1}{k_1}\right] * \dfrac{1}{(A)}} \qquad (5.1)$$

enzyme-catalyzed reaction is defined as the initial velocity at infinite substrate concentration. Therefore, the V_{max} for the chemical mechanism in Figure 5.1 would be the right side of equation 5.1 multiplied by E_t but without the denominator term that is multiplied by 1/A, since it becomes insignificant at infinite substrate concentration (equation 5.2). Of course the k_{cat} is the same with the term for E_t in the numerator replaced by 1.

$$V_{max} = \frac{E_t}{\dfrac{1}{k_5} + \dfrac{1}{K_3 * k_5} + \dfrac{1}{k_3}} \qquad (5.2)$$

Furthermore, the V_{max}/K_A is the first-order rate constant for the reaction as the substrate concentration approaches zero. Therefore, the V_{max}/K_A for the chemical mechanism in Figure 5.1 would be the right side of equation 5.1 multiplied by E_t and divided by (A) but with only the denominator term that is multiplied by 1/A, since it becomes the only significant term in the denominator as the substrate concentration approaches zero (equation 5.3). The k_{cat}/K_A is the same expression with the term for E_t replaced by 1.

$$\frac{V_{max}}{K_A} = \frac{E_t}{\dfrac{1}{K_1*K_3*k_5} + \dfrac{1}{K_1*k_3} + \dfrac{1}{k_1}}$$

(5.3)

The K_A is the V_{max} (equation 5.2) divided by V_{max}/K_A (equation 5.3). Operationally it is the substrate concentration that gives one-half the maximum velocity. It can be seen (equation 5.4)

$$K_A = \frac{\dfrac{1}{K_1*K_3*k_5} + \dfrac{1}{K_1*k_3} + \dfrac{1}{k_1}}{\dfrac{1}{k_5} + \dfrac{1}{K_3*k_5} + \dfrac{1}{k_3}}$$

(5.4)

that K_A is influenced by every rate and equilibrium constant in the mechanism. Although it is frequently regarded as the dissociation constant of the enzyme-substrate complex, it is so only to the extent that the complex formation from free enzyme and free substrate is actually in equilibrium under the experimental conditions of the steady state. It can also be seen that the K_A is conceptually of less utility in kinetic discussions than the V_{max}/K_A.

The mathematical model can be expressed operationally as the a version of the familiar Michaelis-Menten equation and its double reciprocal form equation 5.5.

$$v_i = \frac{1}{\dfrac{1}{V_{max}} + \dfrac{K_A}{V_{max}} * \dfrac{1}{(A)}}$$

(5.5)

$$\frac{1}{v_i} = \frac{K_A}{V_{max}} * \frac{1}{(A)} + \frac{1}{V_{max}}$$

A somewhat more familiar simpler chemical model, at least mathematically if not conceptually, is one in which there is only one enzyme substrate/product intermediate (Figure 5.2). The mathematical model is derived in the same way (equation 5.6) and can also be expressed operationally as the familiar Michaelis-Menten equation.

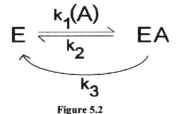

Figure 5.2

$$\frac{v_i}{E_t} = \cfrac{1}{\dfrac{1}{k_3} + \left[\dfrac{1}{k_1} + \dfrac{1}{K_1 * k_3} * \dfrac{1}{(A)} \right]}$$

$$v_i = \cfrac{1}{\dfrac{1}{V_{max}} + \dfrac{K_A}{V_{max}} * \dfrac{1}{(A)}}$$

(5.6)

The corresponding expressions for V_{max}, V_{max}/K_A and K_A (equation 5.7) are determined in the same way as above.

$$V_{max} = k_3 E_t$$

$$\frac{V_{max}}{K_A} = \cfrac{E_t}{\dfrac{1}{k_1} + \dfrac{1}{K_1 * k_3}}$$

$$K_A = \cfrac{\dfrac{1}{k_1} + \dfrac{1}{K_1 * k_3}}{\dfrac{1}{k_3}} = \frac{k_2 + k_3}{k_1}$$

(5.7)

When data is fitted to a model, values for V_{max}, K_A and V_{max}/K_A are obtained. They generally cannot be deconvoluted to obtain values for the individual rate and equilibrium constants in the expressions above. Furthermore, comparison of equation 5.5 with equation 5.6 demonstrates that a chemical mechanism with a single intermediate cannot be distinguished from one with two or more intermediates. Thus one of the limitations of steady-state enzyme kinetic data is demonstrated. Nevertheless each of these parameters has the operational meaning described above. Furthermore, the expressions above demonstrate that V_{max}, and k_{cat}, is sensitive to all of the irreversible steps in a mechanism in addition to all of the contiguous upstream reversible steps except those that are associated with the binding of substrate. The V_{max}/K_A is sensitive to the substrate-binding step in addition to any contiguous, reversible steps and one contiguous downstream irreversible step.

Two of the operational parameters have graphical meaning in the context of the double-reciprocal plot (equation 5.5). The reciprocal of the V_{max}, is the vertical intercept of the double-reciprocal plot and the reciprocal of the V_{max}/K_A, is the slope.

5.3.2. BIREACTANT ENZYMES

Bireactant enzymes present mechanistic issues that make them significantly more interesting that unireactant enzymes. In a particularly useful systematic experimental approach the initial velocity is measured in experiments in which one substrate is held constant and the other substrate is varied. The concentration of the constant substrate is changed and the experiments are repeated. The series is repeated at different concentrations of the constant substrate ideally until there are data with at least five concentrations of each substrate in a range from one-fifth the K_M to about five times the K_M for that substrate. The first substrate is called the variable substrate whereas the other is called the constant, variable substrate. The result is a family of curves of initial velocity as a function of the variable substrate with each curve representing a different concentration of the constant, variable substrate. The double-reciprocal plot should be a family of straight lines each at a different concentration of the constant, variable substrate. Of course the designation of which substrate is the variable and which the constant, variable substrate is purely arbitrary.

The mechanistic issues that can be decided with these experiments generally have to do with the relative order of substrate binding and product release, although the exact order of substrate addition can be distinguished only in special cases.

Sequential, Ordered Substrate Addition
The first chemical model to be discussed and compared with others is that in which both substrates bind to the enzyme in order and then product is released. It is called a bireactant sequential, ordered model (Figure 5.3). The mathematical model (equation 5.8) is derived by inspection. The initial velocity of this mathematical model contains a term for the reciprocal of the concentration of each of the substrates, a cross product term containing the reciprocal of the concentrations of both substrates, and a term without the concentration of either substrate.

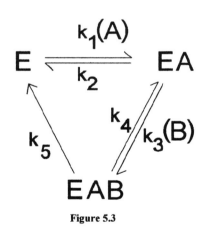

Figure 5.3

$$\frac{v_i}{E_t} = \frac{1}{\dfrac{1}{k_1}*\dfrac{1}{(A)} + \left[\dfrac{1}{K_1*K_3*k_5} + \dfrac{1}{K_1*k_3}\right]*\dfrac{1}{(A)*(B)} + \left[\dfrac{1}{k_3} + \dfrac{1}{K_3*k_5}\right]*\dfrac{1}{(B)} + \dfrac{1}{k_5}} \qquad (5.8)$$

The mathematical model can be simplified somewhat, if combinations of the rate and equilibrium constants are defined in terms of four operational parameters, V_{max}, V_{max}/K_B, $V/_{max}/K_A$, and K_{iA}, and four additional ones calculated from them, k_{cat}, K_B, and K_A. The V_{max} is the initial velocity as both of the substrates approach infinite concentration (equation 5.9).

$$V_{max} = E_t * k_5$$
$$\frac{V_{max}}{K_A} = E_t * k_1$$
$$\frac{V_{max}}{K_B} = \frac{E_t}{\dfrac{1}{k_3} + \dfrac{1}{K_3 * k_5}}$$
$$K_{iA} = K_1$$

(5.9)

The V_{max}/K_A is the first order rate constant as the concentration of A approaches zero, with B approaching infinite concentration. The V_{max}/K_B is the first order rate constant as the concentration of B approaches zero with the concentration of A approaching infinity (equation 5.9). The K_{iA} is the dissociation constant of the enzyme-A complex in the present chemical model (Figure 5.3). Therefore, the operational model for this chemical model is a somewhat simpler version of equation 5.8 (equation 5.10, first equation) and the data from the experiments described above for this chemical model is actually fit to equation 5.10, second equation.

$$v_i = \frac{1}{\dfrac{K_A}{V_{max}} * \dfrac{1}{(A)} + \dfrac{K_{iA} * K_B}{V_{max}} * \dfrac{1}{(A)*(B)} + \dfrac{K_B}{V_{max}} * \dfrac{1}{(B)} + \dfrac{1}{V_{max}}}$$

$$v_i = \frac{(A)*(B)}{\dfrac{K_A}{V_{max}} * (B) + \dfrac{K_{iA} * K_B}{V_{max}} + \dfrac{K_B}{V_{max}} * (A) + \dfrac{1}{V_{max}} * (A)*(B)}$$

(5.10)

Although some of these parameters seem to be rather precisely and unambiguously related to individual rate and equilibrium constants for this chemical model, it will be demonstrated later that the data from other chemical models will fit the same equation just as well. For example it will be shown later that the data for a bisubstrate reaction with random binding of substrates will fit an equation of the same form. Furthermore, even if the reaction is known to proceed with the ordered binding of substrates, there may be reversible steps upstream or downstream from the binding of A, that do not involve the binding of B, there may be additional irreversible steps, *e.g.* dissociation of additional products, or there may be additional intermediates. The rate and equilibrium constants of these additional steps will not affect the form of the equation, but they will certainly affect the meaning and values of the estimated parameters, V_{max}/K_A, K_{iA}, V_{max}/K_B, V_{max}.

Graphically, the double-reciprocal plot with either $1/(A)$ or $1/(B)$ on the horizontal axis will be a family of straight lines that will cross the vertical axis at the points, determined by the concentration of the constant-variable substrate (Figure 5.4). Thus the vertical intercept for the double-reciprocal plot of $1/v_i$ *vs* $1/(A)$ contains a term for the concentration of B (equation. 5.11), which will thus determine the point at which it crosses (Figure 5.4). The presence of a cross product term determines that the family of lines will intersect at the same point, which can be demonstrated from the solution for $1/(A)$ of two

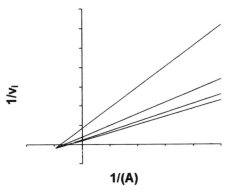

1/(A)

Figure 5.4

$$\frac{1}{v_i} = \frac{1}{\left[\dfrac{K_b}{(B)} + 1\right] * \dfrac{1}{V_{max}}}$$

(5.11)

simultaneous linear equations (*i.e.* the double-reciprocal version of equation 5.10) with two different concentrations of B and substitution of the result back into the original double-reciprocal equation. Thus the intersection point (equation 5.12) of the family of straight lines resulting from the plots of $1/v_i$ *vs* $1/(A)$ contains no term for the concentration of the either substrate.

$$\frac{1}{(A)} = -\frac{1}{K_{iA}}$$

$$\frac{1}{v_i} = -\frac{1}{V_{max}} * \left[1 - \frac{K_A}{K_{iA}}\right]$$

(5.12)

Bisubstrate, sequential, ordered, rapid-equilibrium Substrate Addition.
Frequently some chemical species that act kinetically as substrates of an enzyme are not released at all as products, *e.g.* metal ions. However, before these can be discussed explicitly it is useful to discuss reactions in which the first substrate to bind in a bisubstrate sequential ordered model is actually in rapid equilibrium with the enzyme (Figure 5.5). In such a model the enzyme complex with A, EA, dissociates back to free enzyme, E, and substrate, A, faster than the complex adds the second substrate, B. Thus the magnitude of k_2,

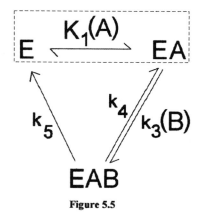

Figure 5.5

implicit in K_1, is greater than that of k_3. The conventions for derivation of models with rapid equilibrium steps, described by Cha [2], are that a rapid equilibrium segment (dashed box in Figure 5.5) can be treated as a single enzyme form, the equation is derived, and then the concentration of the individual species comprising the equilibrium segment are related to the total of the equilibrium segment by its equilibrium constant. The rapid-equilibrium segment will be regarded as a single species with a total concentration of enzyme intermediates E' (equation 5.13). Then the preliminary equation for v_i/E_t is equation 5.14.

$$(E') = (E) + (EA) \tag{5.13}$$

The convention is that any rate constant in the velocity expression of a step leading away from the rapid-equilibrium segment should be multiplied by an expression describing the

$$\frac{v_i}{E_t} = \frac{1}{\left[\dfrac{1}{k_3} + \dfrac{1}{K_3 * k_5}\right] * \dfrac{1}{(B)} + \dfrac{1}{k_5}} \tag{5.14}$$

fraction of the total concentration of the enzyme species in the segment that is the individual subspecies reacting in that step. For example any term in equation 5.14 containing k_3, including K_3, should be multiplied by the fraction of E' that is actually EA. That fraction is expressed by equation 5.15.

$$\frac{(EA)}{(E')} = \frac{(EA)}{(E)+(EA)} \tag{5.15}$$

Substitution of the expression (equation 5.16) for the equilibrium of the dissociation of EA into equation 5.15 yields equation 5.17, for the fractional concentration of EA. The latter can be simplified as described.

$$K_1 = \frac{(E)*(A)}{(EA)} \tag{5.16}$$

$$\frac{(EA)}{(E')} = \frac{(EA)}{(EA)+\dfrac{K_1}{(A)}*(EA)}$$

$$\frac{(EA)}{(E')} = \frac{1}{1+\dfrac{K_1}{(A)}} \tag{5.17}$$

Now every term in equation 5.14 that contains k_3, or K_3, is multiplied by the fractional concentration of EA (equation 5.17) to yield equation 5.18 and the somewhat simplified version, equation 5.19.

$$\frac{v_i}{E_t} = \frac{1}{\left[\dfrac{1}{k_3}+\dfrac{1}{K_3*k_5}\right]*\dfrac{K_1}{(B)*(A)} + \left[\dfrac{1}{k_3}+\dfrac{1}{K_3*k_5}\right]*\dfrac{1}{(B)}+\dfrac{1}{k_5}} \tag{5.18}$$

$$v_i = \frac{1}{\dfrac{K_{iA}*K_B}{V_{max}}*\dfrac{1}{(B)*(A)} + \dfrac{K_B}{V_{max}}*\dfrac{1}{(B)}+\dfrac{1}{V_{max}}} \tag{5.19}$$

Graphically the double-reciprocal plot with the second substrate as the variable substrate and the first substrate as the constant-variable substrate consists of a family of lines that intersect on the vertical axis (Figure 5.6), since the mathematical model contains no term for $1/(A)$ that does not also contain $1/(B)$. The plot with the first substrate as the variable substrate will result in a family of lines that intersect to the left of the vertical axis. Nevertheless, the fits of data from the appropriate experiments to this model should be able to distinguish it from other bisubstrate reaction models.

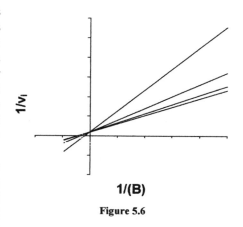

1/(B)

Figure 5.6

For example the enzyme octopine dehydrogenase binds NAD and octopine in that order. The data for the initial velocity at various concentrations of both NAD and octopine fit to equation 5.19 and could be distinguished from other plausible models, *e.g.* equation 5.10 [3]. Furthermore, the double reciprocal plot of $1/v_i$ *vs* $1/(octopine)$ at various concentrations of NAD was a family of lines intersecting on the vertical axis. This analysis supported the hypothesis that NAD binds to the enzyme in a rapid equilibrium manner.

The reader may wish to verify the fact that a chemical model in which the second substrate, B, also binds in rapid-equilibrium manner results in a mathematical model indistinguishable from equation 5.19. In addition a chemical model in which the second substrate binds in rapid-equilibrium manner but the first substrate binds in a steady-state reaction (dissociates more slowly) predicts a mathematical model indistinguishable from equation 5.10.

Any first substrate that is not released from the enzyme during the catalytic cycle (Figure 5.7) will give the same mathematical model and graphical pattern. This has been true most frequently with metal ions. The segment of the model in the dashed box can be regarded as a rapid-equilibrium segment and the mathematical model derived the same way as that above.

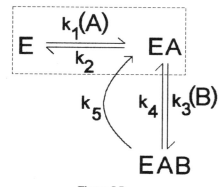

Figure 5.7

For example Jane *et al.* investigated the enzyme phosphoenolpyruvate carboxylase [4]. The initial-velocity data with phosphoenolpyruvate as the variable substrate and Mg^{2+} as the constant-variable substrate fit the mathematical model above (equation 5.19) and graphically the double reciprocal plot was a family of lines that intersected on the vertical axis.

The same model describes the initial velocity of enzymes that react with substrates on insoluble particles, *e.g.* micelles. For example phospholipase A_2 catalyzes the hydrolysis of triglycerides in micelles. Deems *et al.* [5] showed that a similar model describes the initial-velocity data if the physical concentration of the micelles is regarded as the first substrate and the concentration of triglyceride in the micelle is regarded as the concentration of the second substrate. One of the difficulties with this approach to enzymes in insoluble systems is to know and to be able to vary the concentration of the substrate in the particle. The latter authors accomplished it by dilution of the phospholipid in the micelles with another polar lipid. It is interesting to speculate about the possible applicability of this model to enzymes that react with substrates on the surface of other insoluble particles such as liposomes or even the surface of a cell, but in the latter cases it would be necessary to know and control the two-dimensional concentration of substrate on the surface.

It is of further interest to speculate about the possible applicability of this kinetic model to enzymes that catalyze reaction with polymers in the context of the issue of processivity. In regard to a linear polymer it would be necessary to know and be able to control the one-dimensional concentration of the substrate within the polymer. For example a restriction endonuclease or nucleic-acid modifying enzyme reacts only at certain sites on DNA. If one could know and control the frequency of sites along the DNA, the DNA itself would be the first substrate and the frequency of sites would be the second substrate. In this case such a model might provide a test of processivity.

Sequential, Random Substrate Addition
Bireactant chemical and mathematical models with the ordered addition of substrates to an enzyme should be compared with other sequential models in which there is a random addition of substrates. The chemical model in which the substrates are added randomly and in which the rate of substrate dissociation from the enzyme is comparable to the catalytic reaction and the release of products, a steady-state model, (Figure 5.8) is a branching model, with which the present derivation method does not deal in a convenient way. Although, derivation of the mathematical model by a different method (*e.g.* King Altman) can be accomplished, it results in a complex equation that contains second order terms in reciprocal substrate concentration. The second order terms should result in nonlinearity of the double-reciprocal plots, and the model should be distinguishable from an

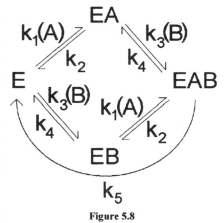

Figure 5.8

ordered model. However, the random, steady-state model is difficult to distinguish because of experimental error and the author knows of no enzyme for which this has actually been accomplished.

The random model more commonly tested is one in which both substrates bind and dissociate at a faster rate than the remainder of the reaction, a rapid-equilibrium random mechanism   (Figure 5.9).   The mathematical model for this chemical model can be derived with the principles of Cha [2] for rapid-equilibrium segments described above. The initial equation for initial velocity contains only one term (equation 5.20), but the equation (equation 5.21) for the fractional distribution of the equilibrium segment in the species EAB is somewhat more complicated. Multiplication of the rate constant, k_5, for the reaction of EAB (to product and E) by the fractional distribution in EAB results in equations 5.22,   one   with   mechanistic parameters, the rate and equilibrium constants

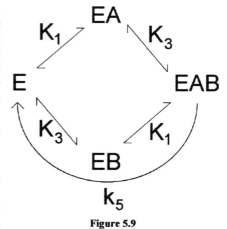

Figure 5.9

$$\frac{v_i}{E_t} = \frac{1}{\dfrac{1}{k_5}}$$

(5.20)

$$(E') = (EAB) + (EB) + (EA) + (E)$$

$$K_1 = \frac{(EA)}{(E)*(A)} = \frac{(EB)}{(EB)*(A)}$$

$$K_3 = \frac{(EB)}{(E)*(B)} = \frac{(B)}{(EA)*(B)}$$

$$f_{EAB} = \frac{(EAB)}{(E')} = \frac{(EAB)}{\dfrac{1}{(A)*K_1}*(EAB) + \dfrac{1}{(B)*K_3}*(EAB) + \dfrac{1}{(A)*(B)*K_1*K_3}*(EAB) + (EAB)}$$

(5.21)

$$f_{EAB} = \frac{1}{\dfrac{1}{(A)*K_1} + \dfrac{1}{(B)*K_3} + \dfrac{1}{(A)*(B)*K_1*K_3} + 1}$$

from the chemical model (Figure 5.9), and one with slightly simpler operational

$$\frac{v_i}{E_t} = \cfrac{1}{\cfrac{1}{K_1*k_5}*\cfrac{1}{(A)} + \cfrac{1}{K_3*k_5}*\cfrac{1}{(B)} + \cfrac{1}{K_1*K_3*k_5}*\cfrac{1}{(A)*(B)} + \cfrac{1}{k_5}}$$

$$v_i = \cfrac{1}{\cfrac{K_{iA}}{V_{max}}*\cfrac{1}{(A)} + \cfrac{K_{iB}}{V_{max}}*\cfrac{1}{(B)} + \cfrac{K_{iA}*K_{iB}}{V_{max}}*\cfrac{1}{(A)*(B)} + \cfrac{1}{V_{max}}}$$

(5.22)

Graphically the mathematical model for a bireactant, sequential, rapid-equilibrium random chemical model describes a double-reciprocal plot consisting of a family of straight lines that intersect the vertical axis at different points (*i.e.* equation 5.23, when A is the variable substrate and B is the constant, variable substrate). Furthermore, they converge at a single point, whose coordinates are in equation 5.24, to the left of the vertical axis regardless whether A or B is the variable substrate. That the convergence point is on the horizontal axis is a necessary but not a unique condition for the chemical model, since the steady-state ordered model (Figure 5.3) will do the same thing if $K_A = K_{iA}$ (equation 5.12).

$$\frac{1}{v_i} = \frac{1}{V_{max}} * \left[\frac{K_{iB}}{(B)} + 1 \right]$$

(5.23)

$$\frac{1}{v_i} = 0$$

$$\frac{1}{(A)} = -\frac{1}{K_{iA}}$$

(5.24)

$$\frac{1}{(B)} = -\frac{1}{K_{iB}}$$

In addition the mathematical model for a bireactant, sequential, rapid-equilibrium random chemical model is generally indistinguishable from that for a bireactant, sequential ordered chemical model (equation 5.9). Theoretically the equation for the ordered model has one more parameter than that for the random model and could be distinguished by curve fitting, but in practice the data is seldom sufficiently precise to make the distinction with satisfactory certainty. Therefore, additional kinds of experiments will be necessary to establish order of binding of substrates to an enzyme.

Examples of random models will be discussed more thoroughly, after methods have been presented that can distinguish them from ordered models.

Nonsequential Substrate Addition
A chemical model that results in a rather distinctive mathematical model for the experiments described above is one in which there is an irreversible step between the addition of the two substrates (Figure 5.10). The mathematical model with the mechanistic parameters (equation 5.25) can also be expressed as a model with operational parameters (equation 5.26), but the operational parameters K_A and V_{max} have a somewhat different meaning (equation 5.27) than in the previous bisubstrate models. Thus, the mechanistic

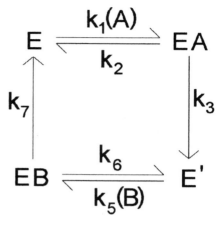

Figure 5.10

$$\frac{v_i}{E_t} = \frac{1}{\left[\dfrac{1}{k_1} + \dfrac{1}{K_1 * k_3}\right] * \dfrac{1}{(A)} + \left[\dfrac{1}{k_5} + \dfrac{1}{K_5 * k_7}\right] * \dfrac{1}{(B)} + \dfrac{1}{k_3} + \dfrac{1}{k_7}} \qquad (5.25)$$

$$v_i = \frac{1}{\dfrac{K_A}{V_{max}} * \dfrac{1}{(A)} + \dfrac{K_B}{V_{max}} * \dfrac{1}{(B)} + \dfrac{1}{V_{max}}} \qquad (5.26)$$

interpretation of operational parameters depends upon the chemical model.

In addition the irreversible step causes the mathematical model (*e.g.* equation 5.26) to have no term containing the reciprocal concentrations of both substrates, (A) and (B). Therefore, data from the experiments described above with an enzyme that complies with this model can generally be distinguished from the sequential bisubstrate models discussed aboveby curve fitting.

$$V_{max} = \frac{E_t}{\dfrac{1}{k_3} + \dfrac{1}{k_7}}$$

$$\frac{V_{max}}{K_A} = \frac{E_t}{\dfrac{1}{k_1} + \dfrac{1}{K_1 * k_3}}$$

(5.27)

$$\frac{V_{max}}{K_B} = \frac{E_t}{\dfrac{1}{k_5} + \dfrac{1}{K_5 * k_7}}$$

The most common cause of an irreversible step between the binding steps for the two substrates is the release of a product before the binding of the second substrate to bind. Under the usual conditions to measure initial velocity there is a negligible concentration of either product and both steps for product release are irreversible. In this model the addition of the two substrates is nonsequential.

Graphically this model results in a double-reciprocal plot of a family of parallel lines rather than the intersecting lines typical of bisubstrate models without the irreversible step (Figure 5.11). Each of the lines intersects the vertical axis at a different point, since the equation for the vertical intercept (e.g. with A as the variable substrate, equation 5.28) contains a term for the concentration of the constant, variable substrate, B. Since the mathematical model is symmetric, the analogous plot with B as the variable substrate will be similar.

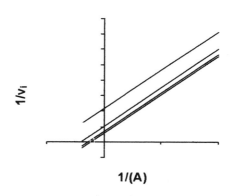

Figure 5.11

$$\frac{1}{v_i} = \frac{K_B}{V_{max}} * \frac{1}{(B)} + \frac{1}{V_{max}}$$

(5.28)

This mechanism is called a double-displacement mechanism or a ping-pong mechanism. The enzyme is in some way changed by the reaction of the first substrate and is then changed back to its original condition by the reaction of the second. Most of the examples of bireactant enzymes that fit this model are transferases, in which the group to be transferred first forms a bond with the enzyme and the remainder of the donor species dissociates as the first product. After the second substrate binds, the group is then transferred to the second substrate in the second half of the reaction. Therefore, the identity of the first substrate to bind to the enzyme is usually unambiguous. However, in the case of some terreactant ligases discussed later there is more than one donor species and the issue requires additional experiments.

Galactose-1-phosphate uridyltransferase catalyzes the transfer of UMP from UTP to galactose-1-phosphate to yield UDP-galactose. Wong and Frey [6] demonstrated that initial velocity data with either UTP or galactose-1-phosphate as the variable substrate fit the double-displacement model and produced a parallel family of lines in the double-reciprocal plot. In addition the initial-velocity data of an abzyme which transfers a phenylacetate group from the vinyl ester to an alcohol, fits the double displacement model [7] and the parallel graphic representation.

If the initial velocity of an enzyme catalyzing a double displacement reaction is determined as a function of substrate concentration as described above, but in the presence of one of the products; the step becomes reversible (Figure 5.12). Now the mathematical model should contain a cross product and the pattern of the double reciprocal plots will become intersecting (equation 5.29). These experiments in the presence of one of the producs are a good confirmation of the mechanistic hypothesis of a double displacement mechanism.

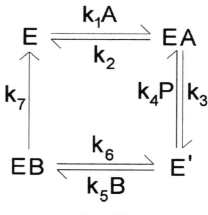

Figure 5.12

In addition there are other confirmatory tests that should be applied. It is usually possible to demonstrate isotopic exchange between one substrate-product pair in the absence of the other pair. Furthermore it is usually possible to demonstrate substrate inhibition by one or both substrates. The latter will be discussed in Chapter 7.

$$\frac{v_i}{E_t} = \frac{1}{D}$$

$$D = \left[\frac{1}{k_1} + \frac{1}{K_1 * k_3}\right] * \frac{1}{(A)} + \left[\frac{1}{K_1 * K_3 * k_5} + \frac{1}{K_1 * K_3 * K_5 * k_7}\right] * \frac{(P)}{(A) * (B)}$$

$$+ \left[\frac{(P)}{K_3 * k_5} + \frac{(P)}{K_3 * K_5 * k_7} + \frac{1}{k_5} + \frac{1}{K_5 * k_7}\right] * \frac{1}{(B)} + \frac{1}{k_3} + \frac{1}{k_7}$$

$$(5.29)$$

5.3.3 TERREACTANT ENZYMES

The same kinds of experiments are done with terreactant enzymes that were done with bireactant enzymes, but with only two substrates at a time and one substrate at a constant concentration. Thus a complete investigation requires three times as many experiments as those required for a bireactant enzyme. The nonvariable substrate in these experiments is ordinarily kept at a concentration approximately equal to its K_M.

The mechanistic possibilities become more numerous and an exhaustive taxonomy of the possibilities becomes somewhat complicated [8]. In addition no enzyme has been shown to fit some of the possible models. Therefore, the approach here will be to discuss some of the models that can be distinguished and comment on some of the general principles involved. Specifically the terreactant sequential ordered model will be discussed and then compared with the models that result as modifications of specific segments in which two substrates bind to the enzyme.

Sequential, Ordered Model
The chemical model for an ordered sequential terreactant, sequential enzyme (Figure 5.13) results in a mathematical model (equation 5.30) that contains terms for each possible binary cross product of substrate concentration except for the cross product of the first and third substrate to bind (*i.e.* A and C). It also contains the cross product of all three substrate concentrations. Therefore, there is no term for the cross product of the first and third substrates to bind that does not also contain the second substrate.

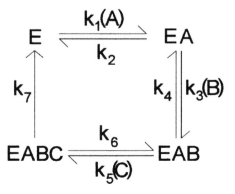

Figure 5.13

$$\frac{v_i}{E_t} = \frac{1}{D}$$

$$D = \frac{1}{k_1} * \frac{1}{(A)} + \frac{1}{K_1 * k_3} * \frac{1}{(A)*(B)} + \frac{1}{k_3} * \frac{1}{(B)}$$

(5.30)

$$+ \left[\frac{1}{K_3 * k_5} + \frac{1}{K_3 * K_5 * k_7}\right] * \frac{1}{(B)*(C)} + \left[\frac{1}{k_5} + \frac{1}{K_5 * k_7}\right] * \frac{1}{(C)}$$

$$+ \left[\frac{1}{K_1 * K_3 * k_5} + \frac{1}{K_1 * K_3 * K_5 * k_7}\right] * \frac{1}{(A)*(B)*(C)} + \frac{1}{k_7}$$

If the binding of the three substrates is ordered and the concentration of the middle substrate to bind is sufficiently high, to render its binding nearly irreversible; the cross product term in the mathematical model for the remaining two substrates (bireactant model) will approach zero and the data will fit best to the equation (equation 5.26) of the double-displacement model described above (Figure 5.10). Furthermore the pattern of the double reciprocal plots will be parallel. However, it may be necessary to do experiments at several concentrations of the putative second substrate and construct plots of the coefficient of the ternary cross product term *vs* the reciprocal of the concentration of the putative second substrate. Extrapolation of the coefficient value to zero at infinite concentration of putative second substrate (the reciprocal equals zero) is evidence that the putative second substrate is indeed the second substrate.

By contrast if either the first two substrates or the last two substrates to bind do so randomly (*e.g.* Figure 5.14), each of the random substrates will cause this shift in models, when at saturating concentration.

The ligase, carbamoyl-phosphate synthetase, catalyzes the reaction of bicarbonate, ammonia and two molecules of MgATP, to yield two MgADP, carbamoyl-phosphate and inorganic phosphate. Raushel *et al.* [9] found that initial-velocity data with all three pairs of substrates fit best to equation 5.10 and the double-reciprocal plot resulted in a family of intersecting lines. The data supported the authors conclusion of a sequential model for the binding of these substrates. When similar experiments with MgATP and ammonia were conducted in the presence of bicarbonate at a concentration of fifteen times its K_M, the data fit best to equation 5.26 and the double-reciprocal plot resulted in a family of parallel lines.

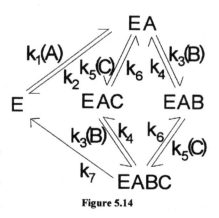

Figure 5.14

The latter data supports the conclusion of an ordered chemical model with bicarbonate as the second substrate to bind. However, the identity of the first and third substrates to bind were established only with additional experiments.

Thus, evidence for the second substrate to bind designates neither the first nor the third substrate to bind. Furthermore, it is theoretically possible for the first and third substrates to bind to do so randomly, but the chemistry becomes somewhat complicated. Further elucidation requires additional experiments.

Other Terreactant Models

In alternative models product may be released from either or both of the enzyme-substrate complexes prior to the binding of the final substrate. In this case initial-velocity experiments in which the variable substrate and the constant-variable substrate bind on either side of, before and after, the product release will fit the equation (equation 5.26) for the double-displacement chemical model (Figure 5.10) even at moderate concentrations of the constant substrate. The double-reciprocal plot of the data should describe a family of parallel lines. However, the data from experiments in which the two substrates are on the same side of, either before or after, the product release will fit the equation (equation 5.10 and/or equation 5.22) for a sequential chemical model. The double-reciprocal plot of the data should describe a family of intersecting lines. However, any two substrates can alternatively bind as a random segment, or an ordered segment. As discussed above additional experiments will be required to determine which.

Asparagine synthetase is a ligase catalyzing the reaction of ATP, aspartate and glutamine to form asparagine, AMP, pyrophosphate and glutamate. Initial-velocity experiments [10] with both substrate pairs including glutamine produced data whose double-reciprocal plot resulted in a family of parallel lines, whereas experiments with ATP and aspartate resulted in intersecting lines. These data supported the authors' conclusions for a chemical model in which the aspartate and ATP constitute a sequential binding segment separated from the binding of glutamine by a product-release step. However, since the released product could be either glutamate or pyrophosphate, the question whether the sequential segment precedes or follows the binding of glutamine required additional experiments to resolve.

5.4. Summary

Initial velocity experiments in which the substrate concentration is varied generally yield information about substrate binding. Experiments with unireactant enzymes results in the estimation of the kinetic parameters, but there is only one possible model for substrate binding.

Experiments with bireactant enzymes generally will distinguish sequential from double-displacement models and a model with ordered substrate binding with the first substrate either in rapid equilibrium or not released at all during the catalytic cycle can be distinguished from those with ordered sequential, steady-state binding of substrates and those with random binding of substrates. However, those with sequential, ordered, steady-state binding cannot generally be distinguished from those with sequential, random binding.

Experiments with terreactant enzymes generally will distinguish the same models as those with bireactant enzymes, but with two substrates at a time. However, a model with sequential, ordered binding of substrates can be distinguished in experiments in which the putative second substrate to bind is at very high concentration.

5.5 References

1. Cleland, W.W. "Multireactant Enzyme Kinetics: Nomenclature, Rate Equations," *Biochim. Biophys. Acta*, 67, 104-137 (1963).

2. Cha, S. "A Simple Method for Derivation of Rate Equations for Enzyme-Catalyzed Reactions under the Rapid Equilibrium Assumption or Combined Assumptions of Equilibrium and Steady State," *J. Biol. Chem.* 243, 820-5 (1968).

3. Schrimsher, J.L and Taylor, K.B. "Octopine Dehydrogenase from *Pecten maximus*: Steady-State Mechanism," *Biochemistry*, 23, 1348-53 (1984).

4. Jane, J.W., O'Leary, M.H. and Cleland, W.W. "A Kinetic Investigation of Phosphoenolpyruvate Carboxylase from *Zea Mays*," *Biochemistry*, 31, 6421-6 (1992).

5. Deems, R.A., Eaton, B.R. and Dennis, E.A. "Kinetic Analysis of Phospholipase A_2 Activity toward Mixed Micelles and Its Implications for the Study of Lipolytic Enzymes," *J. Biol. Chem.* 23, 9013-9020 (1975).

6. Wong, L.-J. And Frey, P.A. "Galactose-1-phosphate Uridylyltransferase: Rate Studies Convirming a Uridylyl-Enzyme Intermediate on the Catalytic Pathway," *Biochemistry*, 13, 3889-94 (1974).

7. Wirsching, P., Ashley, J.A., Benkovic, S.J., Janda, K.D. and Lerner, R.A. "An Unexpectedly Efficient Catalytic Antibody Operating by Ping-Pong and Induced Fit Mechanisms," *Science*, 252, 680-5, (1991).

8. Viola, R.E and Cleland, W.W. "Initial Velocity Analysis for Terreactant Mechanisms," *Methods Enzymol.* 87, 353-366 (1982).

9. Raushel, F.M., Anderson, P.M. and Villafranca, J.J. "Kinetic Mechanism of *Escherichia coli* Carbamoyl-Phosphate Synthetase," *Biochemistry*, 17, 5587-91 (1978).

10. Boehlein, S.K., Stewart, J.D., Walworth, E.S., Thirumoorthy, R., Richards, N.G.J. and Schuster, S.M. "Kinetic Mechanism of *Escherichia coli* Asparagine Synthetase B," *Biochemistry*, 37, 13230-8 (1998).

CHAPTER 6

EFFECTS OF ANALOG INHIBITORS

6.1. Introduction

The objective of the present chapter is to show how the results of initial-velocity experiments with reversible analog inhibitors can be interpreted to provide evidence for specific chemical models for enzyme reactions. Specifically the results of experiments with these inhibitors should provide evidence to distinguish random from ordered binding of substrates as well as to confirm the hypothesis of double-displacement or rapid equilibrium models.

6.1.1. CLASSIFICATION OF INHIBITORS

However, before specific chemical models are discussed it is useful to classify exogenous inhibitors globally in order to define the necessary properties of reversible analog inhibitors. Five kinds of inhibitors are discussed roughly in order of decreasing reversibility.

Rapid, Reversible Inhibitors
Rapid reversible inhibitors bind to and dissociate from the enzyme rapidly compared to the time-span involved in the steady state and the initial velocity measurement. Therefore, the fraction of inhibition does not change significantly during the measurement of initial velocity. Reversible inhibitors also bind to the enzyme with a fairly modest binding energy compared to other types of inhibitors classified here. Since the binding energy of the inhibitor is usually somewhat comparable to that of the substrate, the experimental concentration range of the inhibitor should be within a few orders of magnitude of that of the substrate and, therefore, much larger than that of the enzyme. Therefore, the concentration of free inhibitor is approximately the same as that of total inhibitor and there is no need for a formal equation for the conservation of inhibitor.

Slow-Binding, reversible Inhibitors
Slow-binding inhibitors exert a measurable effect on an enzyme that is comparable to or slower than the rate of measurable catalysis. Therefore, the fraction of inhibition becomes greater during the measurement of initial velocity. Chemically the action of these

inhibitors is frequently associated with a conformation change on the part of the enzyme or, in some cases, the inhibitor. Data from measurements of the time course of inhibition itself is most usefully interpreted in a somewhat different theoretical context and are the subject of a later chapter.

Rapid, Tight-Binding Inhibitors
Tight-binding inhibitors bind to and inhibit the enzyme with a binding energy much greater than the binding of the substrate but not as great as that of irreversible inhibitors. Tight-binding inhibitors are demonstrably reversible. Although the designation depends on several factors discussed further in Chapter 9, generally tight-binding inhibitors have a dissociation constant less than 10^{-8}. Therefore, the experimental concentration of the inhibitor is within two orders of magnitude of the concentration of the enzyme itself, and a significant fraction of the inhibitor is bound to the enzyme. Therefore, the interpretation of data from experiments with these inhibitors requires consideration of the conservation of inhibitor as well as that of the conservation of enzyme. Since this complication necessitates the solution of somewhat complicated equations, such studies to elucidate enzyme mechanisms are somewhat rare. This subject will be dealt with more completely in a later chapter.

Slow- Tight-Binding Inhibitors
Slow- tight-binding inhibitors incorporate both dimensions of complication described for the previous two kinds of inhibitors. Although the experiments are not usually initial-velocity experiments, a later chapter of this book will contain a discussion of them.

Irreversible Inhibitors
Irreversible inhibitors, of course, bind to and inhibit the enzyme irreversibly at least in the time involved in the steady state and the initial velocity measurement. The inhibitor changes the enzyme irreversibly by the formation of a covalent bond with the enzyme or changing the enzyme in some other covalent manner, *e.g.* oxidation. The rate of irreversible inhibition depends on the concentration of enzyme and inhibitor, but eventually all of the enzyme activity will be lost if the inhibitor is present in at least as great a concentration as that of the enzyme.

6.1.2. THE PRESENT CHAPTER

The present chapter is about the rapid reversible inhibitors which are analogs of one of the substrates of the reaction. The inhibitor will have at least some of the same structural features as the substrate to which it is an analog, because it is to bind selectively to the same enzyme form as that substrate. Transition-state analogs continue to be of great theoretical importance in the elucidation of enzyme mechanisms. However, in the present context they function as a good analog inhibitor, if they do not not bind too tightly. In the latter case they are tight-binding inhibitors, described above.. The design and identification of good analog inhibitors for every substrate of the enzyme under investigation frequently requires some chemical insight into the catalytic mechanism of the reaction and considerable ingenuity.

6.2. Experiments and Results

In the most useful experiments the initial velocity is measured as a function of the concentration of each of the substrates, one at a time, in the absence of inhibitor and in the presence of various concentrations of inhibitor. The ideal concentration range of inhibitor would be from about one-fifth to five times the dissociation constant of the inhibitor, defined below. In studies of reactions with multiple substrates the concentration of the constant variable substrate is maintained, at least initially, at a concentration about equal to its respective K_M value in an attempt to prevent minor pathways that might become significant at very high concentration. The data from experiments above result in a family of curves for each substrate each curve at a different concentration of inhibitor, including zero concentration.

6.3. Unireactant Models

Unireactant enzymes will be discussed in the context of three inhibition models, competitive, uncompetitive, and noncompetitive, possibly already familiar to readers from a basic biochemistry course.

6.3.1. COMPETITIVE INHIBITORS

In the competitive model the inhibitor behaves as a true analog of the substrate and binds selectively to the same enzyme form as the substrate (Figure 6.1). Because they bind to the enzyme at the same site, competitive inhibitors generally have structural features in common with the substrate.

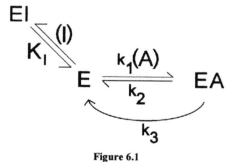

Figure 6.1

       In the derivation of the mathematical model the step in which the inhibitor binds to the enzyme can be treated as a rapid-equilibrium segment, which it is in comparison to the time period during which the initial velocity is measured. Therefore, the sum of the enzyme forms in that segment is designated E_i, and the equation for the initial velocity is written as equation 6.1, which is very similar to equation 5.1 in the previous chapter.

$$\frac{v_i}{E_t} = \frac{1}{\frac{1}{(A)} * \left[\frac{1}{k_1} + \frac{1}{K_1 * k_3}\right] + \frac{1}{k_3}} \tag{6.1}$$

The expression for the fraction of E_i that is free enzyme, f_E, is developed (equation 6.2) as in the previous chapter.

$$K_I = \frac{(I)*(E)}{(EI)}$$

$$f_E = \frac{(E)}{(E_i)} = \frac{(E)}{(E)+(EI)} \tag{6.2}$$

$$f_E = \frac{1}{1 + \frac{(I)}{K_I}}$$

This expression is then multiplied times all of the terms in equation 6.1 that contain k_1 to

$$\frac{v_i}{E_t} = \frac{1}{\frac{1}{(A)} * \left[\frac{1}{k_1} + \frac{1}{K_1 * k_3}\right] * \left[1 + \frac{(I)}{K_I}\right] + \frac{1}{k_3}}$$

$$v_i = \frac{1}{\frac{1}{(A)} * \frac{K_A}{V_{max}} * \left[1 + \frac{(I)}{K_I}\right] + \frac{1}{V_{max}}} \tag{6.3}$$

give equation 6.3, the mathematical model for competitive inhibition, where K_I is the dissociation constant for the enzyme-inhibitor complex and the other parameters have the same meaning as they had in the previous chapter.

Graphically the double-reciprocal plot (Figure 6.2) will describe a family of lines that converge to a point on the vertical axis ($1/(A) = 0$ in equation 6.4). There is a slope effect of different inhibitor concentrations because the coefficient of $1/(A)$ contains a term for inhibitor concentration. There is no intercept effect of inhibitor concentration because the terms that do not contain $1/(A)$, the intercept, also do not have a term for the concentration of inhibitor. Thus competitive inhibitors result in a slope effect but no intercept effect.

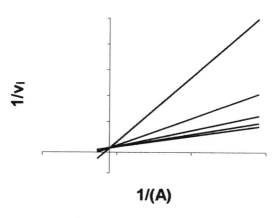

1/(A)

Figure 6.2

$$\frac{1}{v_i} = \frac{1}{(A)} * \frac{K_A}{V_{max}} * \left[1 + \frac{(I)}{K_I}\right] + \frac{1}{V_{max}} \qquad (6.4)$$

6.3.2. UNCOMPETITIVE INHIBITORS

In the uncompetitive chemical model the inhibitor binds selectively to an enzyme form downstream from that to which the substrate has bound (Figure 6.3). Although this uncompetitive inhibitor does not conform to all of the properties of an analog to the substrate, and it is rare in nature for a unireactant enzyme, the model will be useful, when enzymes with multiple substrates are discussed. The mathematical

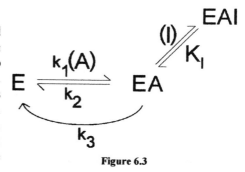

Figure 6.3

model is derived as in the previous example except that the equilibrium segment involves the enzyme-substrate complex, EA, in stead of the free enzyme. In addition the fraction of the equilibrium segment that is EA, f_{EA}, is multiplied times all terms that contain either k_2 or k_3 in equation 6.1 to give equation 6.5. Since $K_1 = k_1/k_2$, f_{EA} appears in the multiplier of $1/(K_1k_3)$ twice, once in the numerator and once in the denominator, they cancel each other.

$$\frac{v_i}{E_t} = \cfrac{1}{\left[\cfrac{1}{(A)} * \left[\cfrac{1}{k_1} + \cfrac{1}{K_1 * k_3} * \cfrac{\left[1 + \cfrac{(I)}{K_I}\right]}{\left[1 + \cfrac{(I)}{K_I}\right]} \right] + \cfrac{1}{k_3} * \left[1 + \cfrac{(I)}{K_I}\right] \right]}$$

(6.5)

$$v_i = \cfrac{1}{\cfrac{1}{(A)} * \cfrac{K_A}{V_{max}} + \cfrac{1}{V_{max}} * \left[1 + \cfrac{(I)}{K_I}\right]}$$

Graphically the double-reciprocal plot (Figure 6.4, equation 6.6) will describe a parallel family of lines that intersect the vertical axis at different points depending on the concentration of the inhibitor. There is no slope effect because the coefficient of $1/(A)$ contains no term for inhibitor concentration. There is an intercept effect because the terms that do not contain $1/(A)$ do contain a term for the concentration of inhibitor. Thus an uncompetitive inhibitor results in an intercept effect but no slope effect.

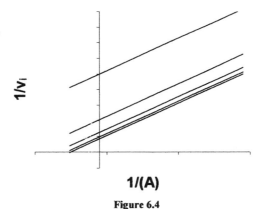

$1/v_i$

$1/(A)$

Figure 6.4

$$\frac{1}{v_i} = \frac{1}{(A)} * \frac{K_A}{V_{max}} + \frac{1}{V_{max}} * \left[1 + \frac{(I)}{K_I}\right]$$

(6.6)

6.3.3. NONCOMPETITIVE INHIBITORS

In the noncmpetitive chemical model the inhibitor binds both to the same enzyme form as the substrate and to the enzyme-substrate complex (Figure 6.5).

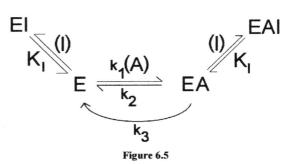

Figure 6.5

Noncompetitive inhibition is also called "mixed inhibition" by some authors.

The mathematical model is derived as in the previous example except that there are two equilibrium segments (equation 6.7). Therefore, every rate and equilibrium constant in equation 6.1 is multiplied times either f_E or f_{EA}, which, in this model, are the

$$\frac{v_i}{E_t} = \cfrac{1}{\left[\cfrac{1}{(A)}*\left[\cfrac{1}{k_1}+\cfrac{1}{K_1*k_3}*\cfrac{\left[1+\cfrac{(I)}{K_I}\right]}{\left[1+\cfrac{(I)}{K_I}\right]}\right]*\left[1+\cfrac{(I)}{K_I}\right]+\cfrac{1}{k_3}*\left[1+\cfrac{(I)}{K_I}\right]\right]}$$

(6.7)

$$v_i = \cfrac{1}{\cfrac{1}{(A)}*\cfrac{K_A}{V_{max}}*\left[1+\cfrac{(I)}{K_I}\right]+\cfrac{1}{V_{max}}*\left[1+\cfrac{(I)}{K_I}\right]}$$

same expression. In addition the multiplier of the equilibrium constant, K_1, contains this expression in both the numerator and the denominator, which, of course, cancel each other.

Graphically the double-reciprocal plot (Figure 6.6, equation 6.8) will describe a family of lines that intersect at a single point to the left of the vertical axis. Therefore, noncompetitive inhibition results in both a slope effect and an intercept effect. In the present model they intersect on the horizontal axis at $-1/K_A$, but that depends upon the

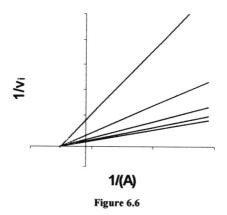

1/(A)

Figure 6.6

fact that the dissociation constant of the EI complex is the same as that of the EAI complex.

$$\frac{1}{v_i} = \frac{1}{(A)}*\frac{K_A}{V_{max}}*\left[1+\frac{(I)}{K_I}\right]+\frac{1}{V_{max}}*\left[1+\frac{(I)}{K_I}\right]$$

(6.8)

Most often the experimental data is fit best by a model in which the two K_I values are different. In the latter case the K_I in the term containing substrate concentration is designated K_{IS}, because it affects the slope, and the other K_I is designated K_{II} because it affects the intercept (equation 6.7 and 6.8).

6.4. Bireactant Models

Experiments with analog inhibitors in bireactant enzyme systems are considerably more interesting and informative than those with unireactant enzymes. The experiments will provide evidence to determine order as well as to confirm some of the models described in the previous chapter. A true analog inhibitor to either substrate A or substrate B will selectively bind to the same enzyme form as its respective substrate. Therefore, there will be a chemical model plus a mathematical model for substrate A and another pair of models for substrate B.

6.4.1. BIREACTANT, SEQUENTIAL MODELS

Three sequential models will be discussed: the ordered model, the rapid-equilibrium, random model and the rapid-equilibrium ordered model.

Sequentia,l ordered model
A bireactant, sequential, ordered system will have two chemical models and two corresponding mathematical models for analog inhibitors, one for the analog to the first substrate to bind and one for the analog to the second substrate to bind. However, there will be four patterns to fit, one for each substrate with the first analog and one for each substrate with the second analog. Although the models below describe the patterns the fitting is actually done with the proper form of the unireactant mathematical models described above. Therefore, the patterns can be described as having a slope effect (competitive), and intercept effect (uncompetitive), or both (noncompetitive).

Analog to the First Substrate. The chemical model for an analog to A, the first substrate to bind (Figure 6.7), results in the mathematical model (equation. 6.9) after derivation by the same algorithm described above for unireactant systems.

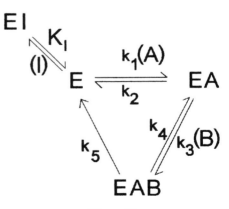

Figure 6.7

$$\frac{v_i}{E_t} = \cfrac{1}{\frac{1}{k_1}*\frac{1}{(A)}*\left[1+\frac{(I)}{K_I}\right]+\left[\frac{1}{K_1*k_3}+\frac{1}{K_1*K_3*k_5}\right]*\frac{1}{(A)*(B)}*\left[1+\frac{(I)}{K_I}\right]+\left[\frac{1}{K_3}+\frac{1}{K_3*k_5}\right]*\frac{1}{(B)}+\frac{1}{k}}$$

(6.9)

$$v_i = \cfrac{1}{\frac{K_A}{V_{max}}*\frac{1}{(A)}*\left[1+\frac{(I)}{K_I}\right]+\frac{K_{iA}*K_B}{V_{max}}*\frac{1}{(A)*(B)}*\left[1+\frac{(I)}{K_I}\right]+\frac{K_B}{V_{max}}*\frac{1}{(B)}+\frac{1}{V_{max}}}$$

Graphically the data from the experiments described above with A as the variable substrate will result in a double-reciprocal plot (equation 6.10) the same as the competitive model described above (Figure 6.2). Furthermore the data will fit the mathematical model for competitive inhibition (equation 6.3) better than the other inhibitor models above.

In contrast the data from the experiments described above with B as the variable substrate will result in a double-reciprocal plot (Figure 6.6, equation 6.10) the same as the noncompetitive model described above. Furthermore the data will fit the mathematical model for noncompetitive inhibition (equation 6.7) better than the other inhibitor models above.

$$\frac{1}{v_i} = \frac{K_A}{V_{max}}*\frac{1}{(A)}*\left[1+\frac{(I)}{K_I}\right]+\frac{K_{iA}*K_B}{V_{max}}*\frac{1}{(A)*(B)}*\left[1+\frac{(I)}{K_I}\right]+\frac{K_B}{V_{max}}*\frac{1}{(B)}+\frac{1}{V_{max}}$$

(6.10)

Analog to the second substrate. The chemical model for an analog to the second substrate to bind (Figure 6.8) results in the mathematical model (equation 6.11) after derivation by the same algorithm's described above for unireactant systems.

Graphically the data from the experiments described above with A as the variable substrate will result in a double-reciprocal plot (equation 6.12) the same as the uncompetitive model described above (Figure 6.4). Furthermore the data will fit the mathematical model for uncompetitive inhibition (equation 6.5) better than the

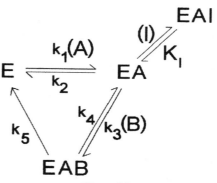

Figure 6.8

$$\frac{v_i}{E_t} = \frac{1}{D}$$

$$D = \frac{1}{(A)} * \frac{1}{k_1} + \frac{1}{(B)*(A)} * \left[\frac{1}{K_1*k_3} + \frac{1}{K_1*K_3*k_5}\right] * \frac{\left[1+\dfrac{(I)}{K_I}\right]}{\left[1+\dfrac{(I)}{K_I}\right]} +$$

$$\frac{1}{(B)} * \left[\frac{1}{k_3} + \frac{1}{K_3*k_5}\right] * \left[1+\frac{(I)}{K_I}\right] + \frac{1}{k_5}$$

$$v_i = \cfrac{1}{\dfrac{1}{(A)} * \dfrac{K_A}{V_{max}} + \dfrac{1}{(B)*(A)} * \dfrac{K_{iA}*K_B}{V_{max}} + \dfrac{1}{(B)} * \dfrac{K_B}{V_{max}} * \left[1+\dfrac{(I)}{K_I}\right] + \dfrac{1}{V_{max}}}$$

$$\frac{1}{v_i} = \frac{1}{(A)} * \frac{K_A}{V_{max}} + \frac{1}{(B)*(A)} * \frac{K_{iA}*K_B}{V_{max}} + \frac{1}{(B)} * \frac{K_B}{V_{max}} * \left[1+\frac{(I)}{K_I}\right] + \frac{1}{V_{max}} \qquad \textbf{(6.12)}$$

other inhibitor models above. This uncompetitive inhibition seen with analogs to the second substrate to bind and the first substrate as the variable substrate is the distinctive evidence for an ordered binding of substrates. It is interesting to repeat the experiments with a high level of the fixed substrate, B in this case, to see if the pattern changes to indicate evidence for a minor random pathway described later.

In contrast the data from the experiments described above with B as the variable substrate will result in a double-reciprocal plot (equation 6.12) the same as the competitive model described above (Figure 6.2). Furthermore the data will fit the mathematical model for competitive inhibition (Eq 6.3) better than the other inhibitor models above. This pattern confirms the fact that the inhibitor is an analog of B.

For example [1] the initial velocity at 25 °C. of cyclopentanol oxidation by the secondary alcohol dehydrogenase from the thermophile, *Thermoanaerobium brockii*, is inhibited by hexafluorocyclopropanol competitively, when cyclopentanol is the variable substrate but uncompetitively, when NADPH is the variable substrate. These patterns constitute evidence that NADPH and cyclopentanol bind to the enzyme in that order. The patterns were unchanged at 60 °C. a more physiological temperature for the enzyme, nor was the uncompetitive pattern changed, when the concentration of cyclopentanol was kept at ten times the K_B.

The inhibition patterns for a bireactant, sequential, ordered model are summarized in a table (Table 6.1)

Sequential, Rapid-Equilibrium, Random Model
The inhibition patterns described by data from inhibition experiments with an enzyme system that fits a bireactant, sequential, random (rapid equilibrium) chemical model (analog to A: Figure 6.9) is significantly different from those of the ordered model.

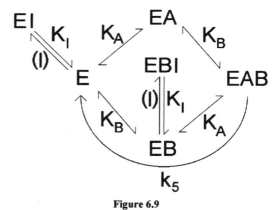

Figure 6.9

The mathematical model for the random chemical model of inhibitor experiments with an analog to A is derived in the same way as that for the the substrate experiments in the previous chapter. The basic initial-velocity equation (equation 6.13) contains k_5, which is multiplied times the fraction of the equilibrium segment that is EAB, f_{EAB}, to yield the

$$\frac{v_i}{E_t} = \frac{1}{\frac{1}{k_5}}$$

$$f_{EAB} = \frac{1}{1 + \frac{K_A}{(A)} + \frac{K_B}{(B)} + \frac{K_A * K_B}{(A)*(B)} + \frac{(I)}{K_I} * \frac{K_A}{(A)} + \frac{(I)}{K_I} * \frac{K_A * K_B}{(A)*(B)}}$$

$$\frac{v_i}{E_t} = \frac{1}{\frac{1}{k_5} * \left[1 + \frac{K_A}{(A)} * \left[1 + \frac{(I)}{K_I}\right] + \frac{K_A * K_B}{(A)*(B)} * \left[1 + \frac{(I)}{K_I}\right] + \frac{K_B}{(B)}\right]}$$

(6.13)

$$v_i = \frac{1}{\frac{K_A}{V_{max}} * \frac{1}{(A)} * \left[1 + \frac{(I)}{K_I}\right] + \frac{K_A * K_B}{V_{max}} * \frac{1}{(A)*(B)} * \left[1 + \frac{(I)}{K_I}\right] + \frac{K_B}{V_{max}} * \frac{1}{(B)} + \frac{1}{V_{max}}}$$

complete initial-velocity expression.
Graphically the data from the experiments described above with A as the variable

substrate will result in a double-reciprocal plot (equation 6.14) that describes a family of lines with the same pattern as the competitive model described above (Figure 6.2). Furthermore the data will fit the mathematical model for competitive inhibition (equation 6.3) better than the other inhibitor models above. This evidence confirms that the inhibitor is a true analog of A.

$$\frac{1}{v_i} = \frac{K_A}{V_{max}} * \frac{1}{(A)} * \left[1 + \frac{(I)}{K_I}\right] + \frac{K_A * K_B}{V_{max}} * \frac{1}{(A)*(B)} * \left[1 + \frac{(I)}{K_I}\right] + \frac{K_B}{V_{max}} * \frac{1}{(B)} + \frac{1}{V_{max}} \qquad (6.14)$$

However, the data from the experiments described above with B as the variable substrate will result in a double-reciprocal plot (equation 6.14) the same as the noncompetitive model described above (Figure 6.6). Furthermore the data will fit the mathematical model for noncompetitive inhibition (equation 6.7) better than the other inhibitor models above.

Since the chemical model is symmetric, the mathematical model for the random chemical model of inhibitor experiments with an analog to B will be the same as that with and analog to A except with all of the A's and B's exchanged for each other and with all of the K_A's and K_B's exchanged for each other (equation 6.15).

$$v_i = \frac{1}{\frac{K_B}{V_{max}} * \frac{1}{(B)} * \left[1 + \frac{(I)}{K_I}\right] + \frac{K_B * K_A}{V_{max}} * \frac{1}{(B)*(A)} * \left[1 + \frac{(I)}{K_I}\right] + \frac{K_A}{V_{max}} * \frac{1}{(A)} + \frac{1}{V_{max}}} \qquad (6.15)$$

In addition the graphic plots and the equations to be fitted to the data will be the same as those with an analog to A with the appropriate exchanges of axis labels and equation terms respectively. With B as the variable substrate both the equation that fits best and the graphic patterns will be competitive, whereas with A as the variable substrate both will be noncompetitive. These patterns are different from those for a bisubstrate, sequential, ordered system (Table 6.1).

For example [2] inhibition of the the initial velocity of 6-phosphogluconate dehydrogenase from *Candida utilis* by 6-sulphogluconate resulted in a competitive pattern, when 6-phosphogluconate was the variable substrate and noncompetitive, when NADP was a variable substrate. The inhibition by ATP-ribose, an analog of NADP, was competitive, when NADP was the variable substrate, and noncompetitive, when 6-phosphogluconate was the variable substrate. These patterns supported the conclusion of a rapid-equilibrium, random binding of NADP and 6-phosphogluconate to the enzyme.

Sequential, Rapid-Equilibrium, Ordered Model
In the previous chapter it was shown that the data from initial-velocity experiments with variable substrate concentrations for a chemical model of a bisubstrate, sequential, ordered system with first substrate in rapid equilibrium could be distinguished from other models of bireactant systems. The fact that the data with analog reversible inhibitors can also be distinguished is used to confirm the basic chemical model for the enzyme system.

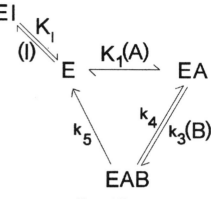

Figure 6.10

Analog to the First Substrate. The chemical model (Figure 6.10) for inhibition with an analog for A results in a mathematical model (equation 6.16), which was derived in the same way as in the previous chapter except that the equilibrium segment and, therefore, the fraction that is EA, f_{EA}, is somewhat different (equation 6.16).

$$\frac{v_i}{E_t} = \frac{1}{\left[\dfrac{1}{k_3} + \dfrac{1}{K_3*k_5}\right]*\dfrac{1}{(B)} + \dfrac{1}{k_5}}$$

$$f_{EA} = \frac{1}{1+\dfrac{K_A}{(A)}+\dfrac{K_A}{(A)}*\dfrac{(I)}{K_I}} = \frac{1}{1+\dfrac{K_A}{(A)}*\left[1+\dfrac{(I)}{K_I}\right]}$$

(6.16)

$$\frac{v_i}{E_t} = \frac{1}{\left[\dfrac{1}{k_3}*\dfrac{1}{K_3*k_5}\right]*\dfrac{1}{(B)} + \left[\dfrac{1}{k_3}*\dfrac{1}{K_3*k_5}\right]*\dfrac{K_A}{(A)*(B)}*\left[1+\dfrac{(I)}{K_I}\right]+\dfrac{1}{k_5}}$$

$$v_i = \frac{1}{\dfrac{K_B}{V_{max}}*\dfrac{1}{(B)}+\dfrac{K_A*K_B}{V_{max}}*\dfrac{1}{(A)*(B)}*\left[1+\dfrac{(I)}{K_I}\right]+\dfrac{1}{V_{max}}}$$

$$\frac{1}{v_i} = \frac{K_B}{V_{max}}*\frac{1}{(B)}+\frac{K_A*K_B}{V_{max}}*\frac{1}{(A)*(B)}*\left[1+\frac{(I)}{K_I}\right]+\frac{1}{V_{max}}$$

(6.17)

Graphically the data from the experiments described above with A as the variable substrate will result in a double-reciprocal plot (equation 6.17.) that describes a family of lines with the same pattern as the competitive model described above (Figure 6.2). Furthermore the data will fit the mathematical model for competitive inhibition (equation 6.3) better than the other inhibitor models above. This evidence confirms that the inhibitor is a true analog of A.

However, the data from the experiments described above with B as the variable substrate will result in a double-reciprocal plot (equation 6.17) that is also the same as the competitive model described above (Figure 6.2). Furthermore the data will fit the mathematical model for competitive inhibition (equation 6.3) better than the other inhibitor models above.

The chemical model (Figure 6.11) for inhibition with an analog for B results in a mathematical model (equation 6.18), which was derived in the same way as the previous model (equation 6.16) except that the equilibrium segment and, therefore, the fraction that is EA, f_{EA}, is somewhat different.

Graphically the data from the experiments described above with A as the variable substrate will result in a double-reciprocal plot (equation 6.19) that describes a family of lines with the same pattern as the uncompetitive model described above (Figure

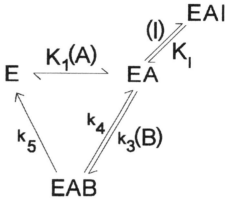

Figure 6.11

$$f_{EA} = \frac{1}{1 + \frac{(I)}{K_I} + \frac{K_A}{(A)}}$$

$$\frac{v_i}{E_t} = \frac{1}{\frac{1}{(B)} * \left[\frac{1}{k_3} + \frac{1}{K_3 * k_5}\right] * \left[1 + \frac{(I)}{K_I} + \frac{K_A}{(A)}\right] + \frac{1}{k_5}}$$ (6.18)

$$v_i = \frac{1}{\frac{K_B}{V_{max}} * \frac{1}{(B)} * \left[1 + \frac{(I)}{K_I}\right] + \frac{K_A * K_B}{V_{max}} * \frac{1}{(A) * (B)} + \frac{1}{V_{max}}}$$

6.4). Furthermore the data will fit the mathematical model for uncompetitive inhibition (equation 6.5) better than the other inhibitor models above. This pattern is analogous to that seen in the ordered, steady-state, sequential, model above.

$$\frac{1}{v_i} = \frac{K_B}{V_{max}} * \frac{1}{(B)} * \left[1 + \frac{(I)}{K_I}\right] + \frac{K_A * K_B}{V_{max}} * \frac{1}{(A)*(B)} + \frac{1}{V_{max}} \tag{6.19}$$

    Graphically the data from the experiments described above with B as the variable substrate will result in a double-reciprocal plot (equation 6.19) that describes a family of lines with the same pattern as the competitive model described above (Figure 6.2). Furthermore the data will fit the mathematical model for competitive inhibition (equation 6.3) better than the other inhibitor models above. This evidence confirms that the inhibitor is a true analog of B.

    Evidence was presented in the previous chapter for the hypothesis that octopine dehydrogenase from *Pecten maximus* binds NAD first in a rapid equilibrium step and then binds octopine [3]. This hypothesis is confirmed by the data describing competitive inhibition by N-ethyl-L-arginine, an analog of octopine, when either octopine or NAD is the variable substrate.

6.4.2. BIREACTANT NONSEQUENTIAL MODEL

The logic of one kind of evidence for the bisubstrate, double-displacement chemical model was described in the previous chapter. The analysis of data from experiments with analog inhibitors provides another kind of evidence.

    The chemical model (Figure 6.12) for inhibition with an analog for A results in a mathematical model (equation 6.20), which was derived with an algorithm similar to that for the ordered, sequential model above.

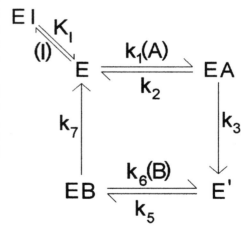

Figure 6.12

$$\frac{v_i}{E_t} = \frac{1}{\frac{1}{(A)}*\left[\frac{1}{k_1}+\frac{1}{K_1*k_3}\right]+\frac{1}{(B)}*\left[\frac{1}{k_5}+\frac{1}{K_5*k_7}\right]+\frac{1}{k_3}+\frac{1}{k_5}}$$

$$\frac{v_i}{E_t} = \frac{1}{\frac{1}{(A)}*\left[\frac{1}{k_1}+\frac{1}{K_1*k_3}\right]*\left[1+\frac{(I)}{K_I}\right]+\frac{1}{(B)}*\left[\frac{1}{k_5}+\frac{1}{K_5*k_7}\right]+\frac{1}{k_3}+\frac{1}{k_5}} \tag{6.20}$$

$$v_i = \frac{1}{\frac{1}{(A)}*\frac{K_A}{V_{max}}*\left[1+\frac{I}{K_I}\right]+\frac{1}{(B)}*\frac{K_B}{V_{max}}+\frac{1}{V_{max}}}$$

Graphically the data from the experiments described above with A as the variable substrate will result in a double-reciprocal plot (equation 6.21) that describes a family of lines with the same pattern as the competitive model described above (Figure 6.2). Furthermore the data will fit the mathematical model for competitive inhibition (equation 6.3) better than the other inhibitor models above. This evidence confirms that the inhibitor is a true analog of A.

$$\frac{1}{v_i} = \frac{1}{(A)}*\frac{K_A}{V_{max}}*\left[1+\frac{I}{K_I}\right]+\frac{1}{(B)}*\frac{K_B}{V_{max}}+\frac{1}{V_{max}} \tag{6.21}$$

The data from the experiments described above with B as the variable substrate will result in a double-reciprocal plot (equation 6.21) that describes a family of lines with the same pattern as the uncompetitive model described above (Figure 6.4). Furthermore the data will fit the mathematical model for uncompetitive inhibition (equation 6.5) better than the other inhibitor models above.

Since the chemical model is symmetric, the mathematical model for the double-displacement chemical model of inhibitor experiments with an analog to B will be the same as that with and analog to A except that all of the A's and B's are exchanged for each other and all of the K_A's and K_B's are exchanged for each other (equation 6.22).

$$\frac{1}{v_i} = \frac{1}{(A)}*\frac{K_A}{V_{max}}+\frac{1}{(B)}*\frac{K_B}{V_{max}}*\left[1+\frac{I}{K_I}\right]+\frac{1}{V_{max}} \tag{6.22}$$

In addition the graphic plots and the equations to be fitted to the data will be the same as those with an analog to A with the appropriate axis labels and equation terms exchanged respectively. With B as the variable substrate both the equation that fits best and the graphic patterns will be competitive, whereas with A as the variable substrate both will be unncompetitive. These patterns (Table 6.1) are different from those for a any of the sequential systems.

The enzyme O-acetylserine sulfhydrylase catalyzes the reaction of O-acetylserine with sulfide ion to form cysteine and acetate. Both initial-velocity experiments with variable substrate and the demonstration of substrate inhibition support the hypothesis of a double-displacement chemical model [4]. Analog inhibition by thiocyanate is competitive, when sulfide is the variable substrte and uncompetitive, when O-acetylserine is the variable substrate. These latter results are consistent with the hypothesized chemical model, although results with an analog of O-acetylserine are not reported.

Sometimes it is difficult to identify a completely selective substrate analog for reactions that conform to the double-displacement model because both substrates may use substantially the same binding site. In the latter case the inhibition patterns will all be noncompetitive.

6.4.3. SUMMARY OF BIREACTANT MODELS

In summary four bisubstrate chemical models can theoretically be distinguished from each other (Table 6.1), steady-state, sequential, ordered; sequential, rapid-equilibrium, random; sequential, ordered with a rapid-equilibrium binding of the first substrate; and steady-state, double-displacement models.

In addition there are several general rules of inhibition that are useful in the discussion of terreactant systems that follows.

1. A substrate analog will inhibit competitively, when the analogous substrate is the variable substrate.

2. A substrate analog will inhibit uncompetitively, when the variable substrate binds to an enzyme form upstream from that to which the inhibitor binds.

3. A substrate analog will inhibit noncompetitively, when the variable substrate binds downstream from the inhibitor, if all of the steps between are reversible. However, if any of the steps is irreversible, the inhibition will be uncompetitive.

Table 6.1: Initial velocity patterns with analog inhibitors in bisubstrate enzyme reactions. C=competitive, UC=uncompetitive, NC=noncompetitive.

Analog to ⇒	A		B	
Variable Substrate ⇒	A	B	A	B
Model				
steady-state, sequential ordered	C	NC	UC	C
rapid-equilibrium, sequential random	C	NC	NC	C
rapid-equilibrium (1st substrate) sequential ordered	C	C	UC	C
double displacement	C	UC	UC	C

6.5. Terreactant Models

In somewhat similar manner to the results with substrate experiments in the previous chapter the results of analog inhibitor experiments with terreactant enzymes can be interpreted by extension of the rules developed for inhibitor experiments with bireactant enzymes. The experiments are very similar to those described above. The initial velocity is measured in the presence of various concentrations of one of the substrates and various concentrations of an analog inhibitor to one of the substrates, while the concentrations of the other two substrates are held constant.

At this point in the overall investigation there is frequently evidence from substrate studies described in the previous chapter whether the best chemical model is completely sequential or whether there is product release, or some other irreversible step, before all of the substrates have bound. Furthermore there may also be preliminary evidence for an ordered, or partially ordered, model from experiments in which the presence of one substrate at very high concentration produced a parallel pattern of the double-reciprocal plot, when the concentrations of the other two substrates were varied.

6.5.1. TERREACTANT, SEQUENTIAL, ORDERED MODEL

In a completely ordered model an analog to the first substrate will be a competitive inhibitor with that substrate but noncompetitive with each of the other two.

An analog to the second substrate in a completely ordered model will be an uncompetitive inhibitor with the first substrate, a competitive inhibitor with the second substrate and a noncompetitive inhibitor with the third substrate.

An analog to the third substrate in the same model will be an uncompetitive inhibitor with each of the first two substrates and a competitive inhibitor with the third substrate.

6.5.2. TERREACTANT, RANDOM, ORDERED MODEL

In a terreactant chemical model with a random segment for first two substrates an analog to either of the two random substrates will be a competitive inhibitor with its analogous substrate and noncompetitive with the other random substrate. For example the chemical model for such a terreactant system in the presence of an analog to A (Figure 6.13) yields a mathematical model (equation 6.23) by the same derivation algorithms used above. Inspection of the latter model reveals that initial-velocity data for a system conforming to

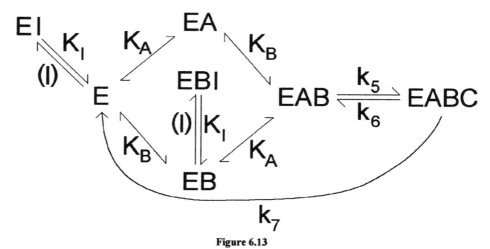

Figure 6.13

this chemical model will be competitive with either A or C as the variable substrate and noncompetitive with B as the variable substrate.

$$\frac{1}{v_i} = \frac{1}{(A)*(C)} * \frac{K_A*K_C}{V_{max}} * \left[1 + \frac{(I)}{K_I}\right] + \frac{1}{(A)*(B)*(C)} * \frac{K_A*K_B*K_C}{V_{max}} * \left[1 + \frac{(I)}{K_I}\right]$$
$$+ \frac{1}{(B)*(C)} * \frac{K_B*K_C}{V_{max}} + \frac{1}{(C)} * \frac{K_C}{V_{max}} + \frac{1}{V_{max}}$$

(6.23)

Since A and B are symmetric in this system, an analog to B will be a competitive inhibitor with either B or C as the variable substrate and noncompetitive with A as the variable substrate. The competitive patterns with C as the variable substrate are analogous to the bireactant, sequential, rapid-equilibrium model above.

In a system that conforms to the same chemical model for substrate binding an analog to the third substrate will be a competitive inhibitor with its analogous substrate, C, and an uncompetitive inhibitor with either of the two random substrates, A or B.

6.5.3. TERREACTANT, ORDERED, RANDOM MODEL

In a similar fashion the inhibition patterns for the system that conforms to a chemical model with an obligatory first substrate and a random segment for the other two can be predicted. The initial velocity will be inhibited competitively by a substrate analog, when the analogous substrate is the variable substrate. The initial velocity will be inhibited noncompetitively by an analog to the first substrate, when either of the other two substrates is the variable substrate. An analog to either of the two random substrates will result in uncompetitive inhibition, when the first substrate is the variable substrate and noncompetitive inhibition, when the other random substrate is the variable substrate.

For example the enzyme octopine dehydrogenase catalyzes the reductive amination of pyruvate by arginine at the expense of NADH to form octopine. The arginine analog, δ-guanidinovalerate, inhibits the initial velocity competitively, when arginine is the variable substrate; noncompetitively, when pyruvate is the variable substrate; and uncompetitively, when NADH is the variable substrate [3]. In addition the pyruvate analog, propionate, inhibits competitively, when pyruvate is the variable substrate; noncompetitively, when arginine is the variable substrate and uncompetitively, when NADH is the variable substrate. These patterns of inhibition provide evidence for the initial binding of NADH followed by random binding of arginine and pyruvate.

6.5.4. TERREACTANT, RANDOM MODEL

The initial velocity of terreactant systems that conform to a completely random chemical model are inhibited competitively by substrate analogs analogous to the variable substrate. However, it is inhibited noncompetitively by analogs analogous to substrates other than the variable substrate. The patterns of this and other models are summarized in Table 6.2.

6.5.5. TERREACTANT, NONSEQUENTIAL MODELS

Systems that conform to chemical models that have one or more product release steps, or other irreversible step, before all of the substrate binding steps are accomplished will usually have distinctive initial-velocity patterns in the substrate experiments described in the previous chapter. Nevertheless, the distinctive patterns produced by inhibition experiments with substrate analogs will provide essential confirmatory evidence.

Initial velocity data that conforms to a chemical model with product release after binding of the first substrate will be inhibited competitively by a substrate analog of the first substrate, A, when A is also the variable substrate. The initial velocity will be inhibited uncompetitively, when either of the other two substrates is the variable substrate, regardless whether their binding is ordered or random.

If the other two substrates bind in random fashion an analog to either of them will result in uncompetitive inhibition, when the first substrate is the variable substrate and noncompetitive inhibition, when the other random substrate is the variable substrate.

If the other two substrates bind in ordered fashion an analog to the second substrate will result in uncompetitive inhibition, when the first substrate is the variable substrate; competitive inhibition, when the second substrate is the variable substrate; and noncompetitive inhibition, when the third substrate is the variable substrate. An analog to the third substrate will result in uncompetitive inhibition, when either the first or second substrate is the variable substrate and competitive inhibition, when the third, analogous, substrate is the variable substrate.

For example asparagine synthetase is a ligase catalyzing the reaction of ATP, aspartate and glutamine to form asparagine, AMP, pyrophosphate and glutamate. Initial-velocity experiments [5] with all combinations of substrates and analog inhibitors to each showed competitive inhibition by each analog with the analogous variable substrate. The glutamine analog (L-glutamic acid γ-methyl ester) inhibited unncompetitively, with either of the nonanalogous variable substrates, and the ATP analog (AMP-PNP) inhibits uncompetitively when glutamine is the variable substrate. These data supported the author's conclusions from the variable substrate studies, described in the previous chapter, for a chemical model in which the aspartate and ATP constitute a sequential binding segment separated from the binding of glutamine by a product-release step. However, since the released product could be either glutamate or pyrophosphate, the question whether the sequential segment preceeds or follows the binding of glutamine required additional experiments to resolve.

By application of the same general principles the patterns of inhibition for a chemical model in which there is a product release step following the binding of the first two substrates as well as for one in which there is product release after the first substrate to bind and after the second (Table 6.2).

In summary a substantial number of the possible chemical models of terreactant systems can be distinguished by interpretation of data from initial-velocity experiments with reversible inhibitors that are substrate analogs. However, the logical elimination of all the alternative models with these experiments alone may require experiments with each possible substrate analog with each possible variable substrate or the determination of nine separate patterns.

Table 6.2: Initial velocity patterns with analog inhibitors in terreactant enzyme reactions. C=competitive, UC=uncompetitive, NC=noncompetitive.

Analog to ⇒	A			B			C		
Variable Substrate ⇒	A	B	C	A	B	C	A	B	C
Model									
sequential ordered	C	NC	NC	UC	C	NC	UC	UC	C
sequential, A & B random, C ordered	C	NC	C	NC	C	C	UC	UC	C
sequential, A ordered, B & C random	C	NC	NC	UC	C	NC	UC	NC	C
sequential, random	C	NC	NC	NC	C	NC	NC	NC	C
product release after A, B & C ordered	C	UC	UC	UC	C	NC	UC	UC	C
product release after A, B & C random	C	UC	UC	UC	C	NC	UC	NC	C
product release before C, A & B ordered	C	NC	UC	UC	C	UC	UC	UC	C
product release before C, A & B random	C	NC	NC	UC	C	UC	UC	UC	C
product release after A and before C	C	UC	UC	UC	C	UC	UC	UC	C

6.6. Summary

In general experiments with the inhibition of initial velocity by substrate analogs is a useful method by which the best steady-state chemical model for an enzyme system can be established. The ordered binding of substrates can be distinguished from random binding, as well as various combinations of the two, in terreactant systems. In addition the method is useful to distinguish chemical models in which product is released before the binding of all of the substrate species. Of course these correlations may be complicated by phenomena such as minor pathways, analogs that overlap both binding sites, and unphysiological conformation changes of the enzyme.

6.6. References

1. Ford, J.B., Askins, J. and Taylor, K.B. "Kinetic Models for Synthesis by a Thermophilic Alcohol Dehydrogenase," *Biotechnol. Bioeng.* 42, 367-75 (1993).

2.Berdis, A.J. and Cook. P.F. "Overall Kinetic Mechanism of 6-Phosphogluconate Dehydrogenase from *Candida utilis*," *Biochemistry*, 32, 2036-40 (1993).

3. Schrimsher, J.L. and Taylor, K.B. "Octopine Dehydrogenase from *Pecten maximus*: Steady-State Mechanism," *Biochemistry*, 23, 1348-53 (1984).

4. Tai, C.-H., Nalabolu, S.R., Jacobson, T.M., Minter, D. and Cook, P.F. "Kinetic Mechanisms of the A and B Isozymes of O-Acetylserine Sulfhydrylase from *Salmonella typhimurium* LT-2 Using the Natural and Alternative Reactants," *Biochemistry*, 32, 6433-42 (1993).

5. Boehlein, S.K., Stewart, J.D., Walworth, E.S., Thirumoorthy, R., Richards, N.G.J. and Schuster, S.M. "Kinetic Mechanism of *Escherichia coli* Asparagine Synthetase B," *Biochemistry*, 37, 13230-8 (1998).

CHAPTER 7

EFFECTS OF PRODUCT INHIBITORS

7.1. Introduction

The presence of the product of an enzyme-catalyzed reaction will inhibit the initial velocity of the reaction. Obviously product inhibition can be studied only in reactions in which there is more than one product because otherwise the presence of a product would violate the conditions for the measurement of initial velocity. The concentration of at least one product must be zero in order to prevent any overall reverse reaction.

The study of product inhibition is rather more frequent than that of analog inhibition, because the identity of the product is more obvious than the identity of a good analog inhibitor. However, the interpretation of product inhibition is somewhat more complex than that of analog inhibition. This additional complexity limits somewhat the usefulness of this approach for the elucidation of steady-state mechanisms unless a thorough study is performed. For example the formation of nonproductive complexes, containing a product molecule, that are not part of the reaction pathway, seen with a number of enzymes, complicates the patterns of product inhibition. These complications will be discussed more fully in the following chapter. In addition enzymes that undergo a rate-limiting isomerization following the release of a product will produce somewhat different inhibition patterns in the presence of that product than do those enzymes that do not undergo such a step, although this distinction will not affect the results of other initial-velocity experiments. Very few enzymes that undergo this isomerization have been described, probably because many investigators are not sensitive to the possibility of distinctive results. Although a slightly simplified version will be presented here the reader is referred to the article by Rudolph [1] for a more complete treatment.

7.2. Experiments and Results

Experimentally the initial velocity is measured in the presence of various concentrations of one of the substrates and various concentrations of one of the products, including zero. Therefore, the results consist of a family of curves of initial velocity *vs* substrate concentration, each at a different concentration of product. The analogous experiment should be repeated for each possible substrate and each possible product.

The data is then fit to the equations derived in the previous chapter, 6.3, 6.5, and

6.7 to determine whether the inhibition pattern is competitive, uncompetitive or noncompetitive respectively. The data can also be plotted, usually as the double-reciprocal plot to see a preliminary indication of the pattern. As will be seen later some of the curves in the double-reciprocal plot may be nonlinear.

The data from each inhibitor level should also be fit to the Michaelis-Menten equation or plotted as the double-reciprocal plot in order to determine the values of the slope and intercept. The slope and intercept should each be plotted, or fitted to a first and second order polynomial *vs* the inhibitor concentration, to determine whether the slope and intercept are each linear or second order; since some of the possible chemical models may result in nonlinear, usually parabolic, inhibition with respect to one, the other or both. The formation of a nonproductive complex of the inhibitor in addition to its catalytic complex will sometimes result in nonlinear inhibition. These will be discussed in greater detail in the following chapter.

7.3. Unireactant Models

The application of product inhibition to unireactant systems is limited, because many of the enzymes, *e.g.* isomerases, have only a single product. However, some unireactant enzymes, *e.g.* lyases, split the substrate into two products, one of which can be present in each series of experiments. The hydrolases also fall in this category because the substrate water is generally not one whose concentration can be varied.

7.3.1. ORDERED PRODUCT RELEASE

The chemical model for the initial velocity of a unireactant enzyme with the ordered release of two products and in the presence of the last one to be released (Figure 7.1) results in the mathematical model (equation 7.1) by the same algorithm for derivation used previously. The product terms appear in the numerator of the denominator as inhibitor terms.

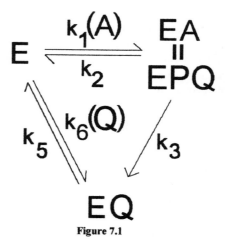

Figure 7.1

$$\frac{v_i}{E_t} = \frac{1}{\frac{1}{(A)}*\left[\frac{1}{k_1}+\frac{1}{K_1*k_3}+\frac{(Q)}{K_5*k_1}+\frac{(Q)}{K_5*K_1*k_3}\right]+\frac{1}{k_3}+\frac{1}{k_5}}$$

$$v_i = \frac{1}{\frac{1}{(A)}*\frac{K_A}{V_{max}}*\left[1+\frac{(Q)}{K_Q}\right]+\frac{1}{V_{max}}}$$

(7.1)

Graphically the double-reciprocal plot of data from an enzyme that conforms to this chemical model is a family of lines that intersects on the vertical axis and is, thus, designated as competitive. The data fits best to the competitive mathematical model (equation 6.3) derived in the previous chapter. The competitive pattern is intuitively more understandable, when one realizes that the substrate and the last product released both bind to the same enzyme form.

The chemical model for the initial velocity of a unireactant enzyme with the ordered release of two products and in the presence of the first one to be released (Figure 7.2) results in the mathematical model (equation 7.2) in which the product terms appear in the numerator of the denominator as inhibitor terms.

Graphically the double-reciprocal plot of data from an enzyme that conforms to this chemical model is a family of lines that intersect to the left of the vertical axis and is, thus, designated as noncompetitive. The data fits best to the noncompetitive mathematical model (equation 6.7) derived in the previous chapter.

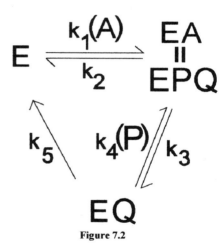

Figure 7.2

$$\frac{v_i}{E_t} = \frac{1}{\frac{1}{(A)}*\left[\frac{1}{k_1}+\frac{1}{K_1*k_3}+\frac{(P)}{K_1*K_3*k_5}\right]+\frac{1}{k_3}+\frac{(P)}{K_3*k_5}+\frac{1}{k_5}}$$

(7.2)

7.3.2. RANDOM PRODUCT RELEASE

In contrast to ordered release of products the chemical model for the initial velocity of a unireactant enzyme with the release of two products in random (rapid equilibrium) fashion and in the presence of one of them (Figure 7.3) results in the mathematical model (equation 7.3) by derivation according to the same algorithm used previously.[1]

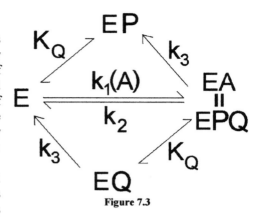

Figure 7.3

Graphically the double-reciprocal plot of data from an enzyme that conforms to this chemical model is a family of lines that intersect on the vertical axis and is, thus, designated as competitive. The data fits best to the competitive mathematical model (equation 6.3) derived in the previous chapter. Since the chemical model is symmetric with respect to product release the same pattern of initial-velocity data will result in the presence of the other product.

The experiments described above can provide evidence to distinguish random from

$$\frac{v_i}{E_t} = \frac{1}{\frac{1}{(A)}*\left[\frac{1}{k_1}*\left[1+\frac{(P)}{K_P}\right]+\frac{1}{K_1*k_3}*\frac{(P)}{K_P}\right]+\frac{1}{k_3}} \tag{7.3}$$

ordered release of products in these unireactant systems as well as the identification of the order to release.

7.4. Bireactant Models

The same principles can be applied to bireactant systems with similar results. However, the possible experiments are somewhat more numerous. With an enzyme that has two substrates and two products there are four sets of experiments. The concentration of each of the substrates is varied in the presence of each of the products. Four different bireactant enzyme models will be discussed: two bireactant, sequential, ordered models (steady-state

---

[1] It was not necessary to account for the rapid-equilibrium partitioning between EPQ and EQ in the rate constant k_3, because the flux through the steps with that rate constant is the sum of both enzyme forms EPQ and EQ.

and rapid equilibrium); the bireactant, sequential, random, rapid equilibriium model and the bireactant nonsequential model.

7.4.1. BIREACTANT, SEQUENTIAL, ORDERED MODELS

The chemical model for the initial velocity of a bireactant enzyme with ordered binding of substrates and the ordered release of two products in the presence of the last product to be released (Figure 7.4) results in the mathematical model in equation 7.4.

When the first substrate to bind, A, is the variable substrate, the initial-velocity data will fit best to a competitive mathematical model (equation 6.3) and the double-reciprocal plot will conform to a family of lines that intersect in a single point on the vertical axis. This competitive pattern is intuitively more understandable, when it is realized that both the product inhibitor and the substrate bind to the same enzyme form.

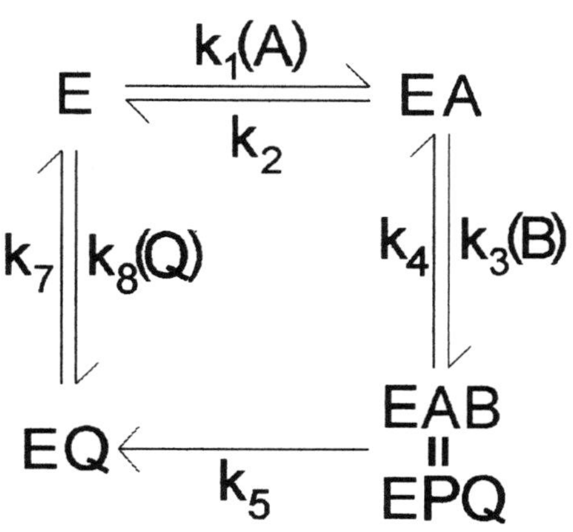

$$E \underset{k_2}{\overset{k_1(A)}{\rightleftharpoons}} EA$$

$$k_7 \Big\Updownarrow k_8(Q) \qquad\qquad k_4 \Big\Updownarrow k_3(B)$$

$$EQ \xleftarrow{\quad k_5 \quad} \begin{matrix} EAB \\ \text{II} \\ EPQ \end{matrix}$$

Figure 7.4

$$\frac{v_i}{E_t} = \frac{1}{D}$$

$$D = \frac{1}{(A)} * \left[\frac{1}{k_1} + \frac{(Q)}{K_7 * k_1} \right] + \frac{1}{(A)*(B)} * \left[\frac{1}{K_1 * k_3} + \frac{1}{K_1 * K_3 * k_5} + \frac{(Q)}{K_7 * K_1 * k_3} + \frac{(Q)}{K_7 * K_1 * K_3 * k_5} \right] +$$

$$\frac{1}{(B)} * \left[\frac{1}{k_3} + \frac{1}{K_3 * k_5} \right] + \frac{1}{k_5} + \frac{1}{k_7} \tag{7.4}$$

$$v_i = \cfrac{1}{\cfrac{1}{(A)} * \cfrac{K_A}{V_{max}} * \left[1 + \cfrac{(Q)}{K_Q} \right] + \cfrac{1}{(A)*(B)} * \cfrac{K_{iA} * K_B}{V_{max}} * \left[1 + \cfrac{(Q)}{K_Q} \right] + \cfrac{1}{(B)} * \cfrac{K_B}{V_{max}} + \cfrac{1}{V_{max}}}$$

When the second substrate to bind, B, is the variable substrate, the initial-velocity data will fit best to a noncompetitive mathematical model (equation 6.7) and the double-reciprocal plot will conform to a family of lines that intersect in a single point to the left of the vertical axis.

Initial-velocity experiments with the same enzyme in the presence of the first product to be released (Figure 7.5 and equation 7.5) and with either substrate as the variable substrate will result in data that fits best to a noncompetitive mathematical model (Eq. 6.7) and a double-reciprocal plot that fits a family of lines that intersect in a single point to the left of the vertical axis.

However, data from experiments with the first substrate as the variable substrate and sufficiently high constant concentrations of the second substrate will conform somewhat better to an uncompetitive pattern.

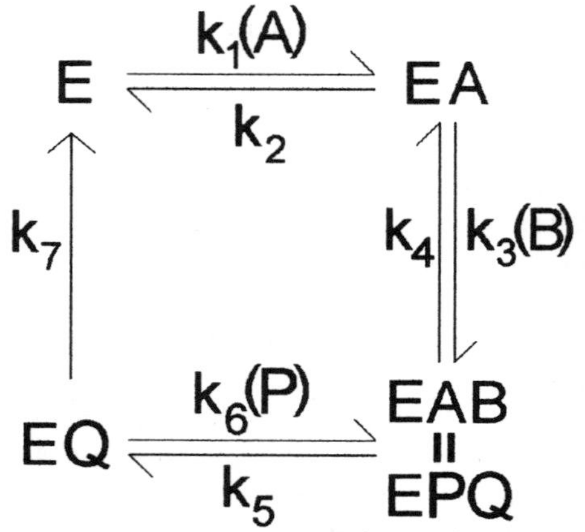

$$E \underset{k_2}{\overset{k_1(A)}{\rightleftharpoons}} EA$$

$$k_7 \uparrow \qquad \qquad k_4 \updownarrow k_3(B)$$

$$EQ \underset{k_5}{\overset{k_6(P)}{\rightleftharpoons}} \begin{array}{c} EAB \\ \text{II} \\ EPQ \end{array}$$

Figure 7.5

$$\frac{v_1}{E_t} = \frac{1}{D}$$

$$D = \frac{1}{(A)} * \frac{1}{k_1} + \frac{1}{(A)*(B)}\left[\frac{1}{K_1*k_3} + \frac{1}{K_1*K_3*k_5} + \frac{(P)}{K_1*K_3*K_5*k_7}\right] + \tag{7.5}$$

$$\frac{1}{(B)}*\left[\frac{1}{k_3} + \frac{1}{K_3*k_5} + \frac{(P)}{K_3*K_5*k_7}\right] + \frac{1}{k_5} + \frac{(P)}{K_5*k_7} + \frac{1}{k_7}$$

The enzyme hypoxanthine-guanine phosphoribosyltransferase, HGPRTase, catalyzes the transfer of ribosylphosphate from phosphoribosylpyrophosphate, PRPP, to either hypoxanthine or guanine to yield inorganic pyrophosphate and either IMP or GMP respectively (Figure 7.6). Product inhibition by IMP was competitive, when PRPP was the variable substrate but noncompetitive, when hypoxanthine was the variable substrate [2]. In addition the inhibition by inorganic pyrophosphate was noncompetitive when either

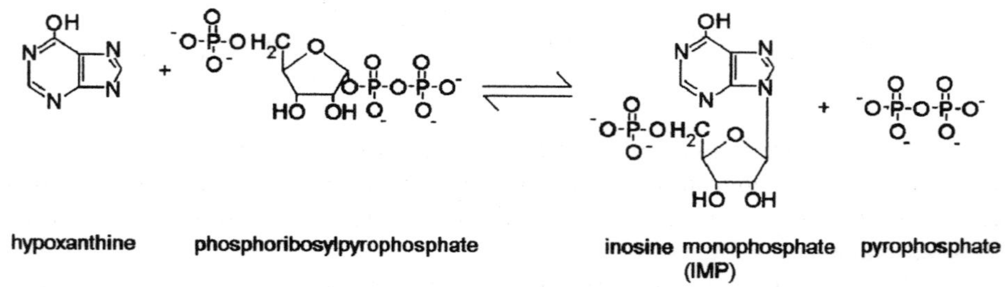

hypoxanthine   phosphoribosylpyrophosphate   inosine monophosphate   pyrophosphate
(IMP)

Figure 7.6

PRPP or hypoxanthine was the variable substrate. These inhibtion patterns supported the investigators' hypothesis that PRPP and hypoxanthine bind to the enzyme in that order and that inorganic pyrophosphate and IMP are released in that order.

Bireactant, Sequential, Ordered, Rapid-Equilibrium Model
In partial analogy to the previous chemical model and in partial analogy to the analog inhibition the chemical model (Figure 7.7) for a bireactant, sequential, reaction with ordered substrate binding and product release and with the first substrate in rapid equilibrium results in a mathematical model that predicts competitive inhibition in the presence of the last product released, when the first substrate is the variable substrate. However, in distinction to the analogous steady-state model just discussed the presence of the same product also results in competitive inhibition, when the second substrate is the variable substrate (equation 7.6).

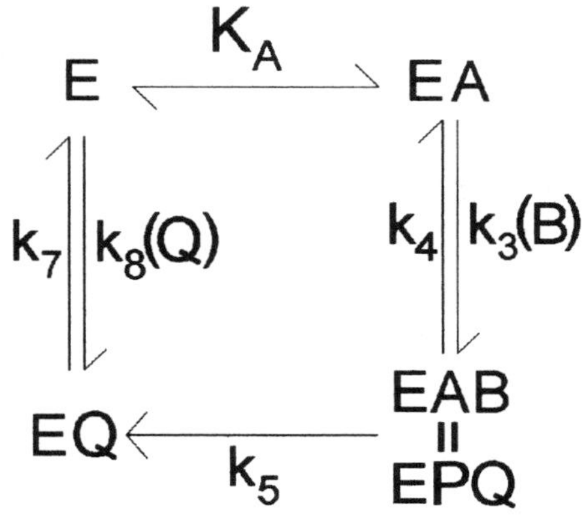

The presence of the first product

Figure 7.7

$$\frac{v_i}{E_t} = \frac{1}{D}$$

$$D = \frac{1}{(B)} * \frac{K_A}{(A)} * \left[\frac{(Q)}{K_7 * k_3} + \frac{(Q)}{K_7 * K_3 * k_5}\right] + \frac{1}{(B)} * \left[\frac{1}{k_3} + \frac{1}{K_3 * k_5}\right] * \left[1 + \frac{K_A}{(A)}\right] + \frac{1}{k_5} + \frac{1}{k_7} \tag{7.6}$$

released, P, results in noncompetitive inhibition when either the first or the second substrate is the variable substrate.

7.4.2. BIREACTANT, SEQUENTIAL, RANDOM MODEL

The discussion of the product inhibition in random models is somewhat more complicated than that for other models. The substrate binding can be random, the product release can be random or both may be random. Thus the chemical models have at least three dimensions of complication, but rather few enzymes have been described to conform to the first two of these chemical models. Furthermore the evidence for them usually derives from experiments in addition to those from initial-velocity kinetics. However, the models will be considered. Furthermore, chemical models with random substrate binding, product release or both may be either steady state or rapid equilibrium. As in the previous chapter the discussion here will be confined to the rapid-equilibrium random chemical model.

The chemical model (Figure 7.8) for product inhibition of a bireactant sequential,

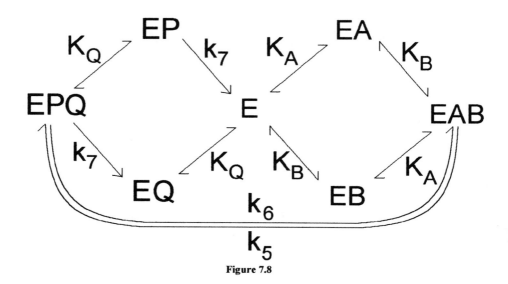

Figure 7.8

reaction that binds substrate and releases product both in a random, rapid-equilibrium mechanism, in which the rate constant for catalysis is much smaller that the rate for the release of product ($k_5 \ll k_7$), results in a mathematical model (equation 7.7) by which competitive inhibition patterns are predicted for both substrates. In addition since the product release is symmetric as well, the inhibition patterns associated with the other product will also be competitive.[2]

$$\frac{v_i}{E_t} = \frac{1}{\frac{1}{k_5} * \left[1 + \frac{K_A}{(A)} + \frac{K_B}{(B)} + \frac{K_A * K_B}{(A) * (B)} * \left[1 + \frac{(Q)}{K_Q} \right] \right]} \qquad (7.7)$$

The enzyme formaldehyde dehydrogenase catalyzes the oxidation of formaldehyde

---

[2] In the original equation there is an additional term containing $1/(K_5 * k_7)$ times an expression containing product concentration that will be between zero and 1.0, but this term will be vanishingly small compared to $1/k_5$.

as the thiohemiacetal with reduced glutathione by the reduction of NAD. It will also catalyze the oxidation of 12-hydroxydodecanoic acid to the ketone. Inhibition by NADH is competitive when either NAD or S-hydroxymethylglutathione is the variable substrate, and inhibition by the keto-dodecanoic acid is competitive, when either NAD or the hydroxydodecanoic acid is the variable substrate [3]. These patterns of competitive inhibition support the investigators hypothesis of a rapid-equilibrium, random binding of substrate and release of products.

Unfortunately the predicted patterns are confused somewhat by the tendency of these enzymes to form dead-end complexes with one substrate and one product. The effect of the formation of such a complex is to change the competitive pattern into a noncompetitive one for the variable substrate with which the complex is formed. Such complexes form most commonly with the substrate-product pair that do not contain the group that is transferred, although the formation of the complementary complex in addition may give rise to two noncompetitive patterns.

Since the discussion here is limited somewhat, the reader is referred to more specialized articles [1],[4] for further considerations.

7.4.3. BIREACTANT, NONSEQUENTIAL MODELS

The chemical model (Figure 7.9) for product inhibition of a bireactant double displacement mechanism results in the mathematical model in equation 7.8. The data from initial-velocity experiments in which the concentrations of both members of a substrate-product pair are varied will fit best to a noncompetitive model (equation 6.3), whereas those in the presence of the same product and the other substrate will fit best to a competitive model (equation 6.7). Since the chemical model is symmetric, experiments in the presence of the other product will produce symmetric results. These patterns are intuitively more understandable when one realizes that a

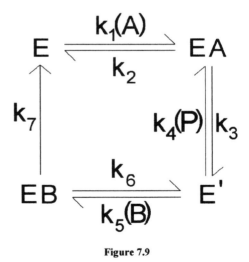

Figure 7.9

substrate-product pair bind to different enzyme forms, whereas the same substrate with the other product bind to the same enzyme form.

$$\frac{v_i}{E_t} = \frac{1}{D}$$

$$D = \frac{1}{(A)} * \left[\frac{1}{k_1} + \frac{1}{K_1 * k_3} \right] + \frac{1}{(A)*(B)} * \left[\frac{(P)}{K_1 * K_3 * k_5} + \frac{(P)}{K_1 * K_3 * K_5 * k_7} \right] +$$

$$\frac{1}{(B)} * \left[\frac{1}{k_5} + \frac{1}{K_5 * k_7} + \frac{(P)}{K_3 * k_5} + \frac{(P)}{K_3 * K_5 * k_7} \right] + \frac{1}{k_3} + \frac{1}{k_7}$$

(7.8)

For example evidence from experiments in which the initial velocity was measured as a function of the substrate concentrations was summarized in Chapter 5 to show that the enzyme galactose-1-phosphate uridylyltransferase (Figure 7.10) conformed to a double

UDP-glucose   UDP-galactose

\+   \+

galactose-1-phosphate   glucose-1-phosphate

Figure 7.10

displacement chemical model with a uridylyl-enzyme intermediate. The results of product inhibition experiments [5] confirmed this hypothesis by the demonstration that inhibition by UDP-galactose was competitive, when UDP-glucose was the variable substrate and noncompetitive, when galactose-1-phosphate was the variable substrate. In the reverse direction galactose-1-phosphate inhibited competitively, when glucose-1-phosphate was the variable substrate and noncompetitively, when UDP-galactose was the substrate. Furthermore, UDP-glucose was a competitive inhibitor, when glucose-1-phosphate was the variable substrate.

However, there are multisite enzymes whose reaction kinetics conforms to a double-displacement model. The latter move the group to be transferred from one site to the other. The product-inhibition patterns with multisite, double-displacement models are complicated by the fact that one or more of the products may form complexes with the enzyme at more than one of the sites leading to additional noncompetitive patterns and nonlinear inhibition. Furthermore, the failure of some products to cause the accumulation of catalytic central complexes may result in uncompetitive inhibition, with respect to one or more substrates. In the interest of simplicity it is sufficient for the investigator to be sensitive to the fact that product-inhibition experiments may not confirm an hypothesis of a double-displacement model for which there is other evidence, but can provide additional interesting mechanistic information. The reader is referred to more detailed discussions for these exceptions.

7.4.4. SUMMARY OF BIREACTANT MODELS

The patterns for product inhibition in the principal chemical models of bireactant systems is summarized in Table 7.1.

Table 7.1: Initial velocity patterns with product inhibitors in bisubstrate enzyme reactions. C=competitive, NC=noncompetitive.

Product ⇒	P		Q	
Variable Substrate ⇒	A	B	A	B
Model				
steady-state, sequential ordered	NC	NC	C	NC
rapid-equilibrium, sequential random	C	C	C	C
rapid-equilibrium (1st substrate) sequential ordered	NC	NC	C	C
double displacement	NC	C	C	NC

Rules for Bireactant Models

Several general rules about product inhibition patterns are useful. However, since there are a number of opportunities for exceptions to the rules, it is usually necessary to derive the mathematical model from the individual chemical model under consideration.

1. If the product binds only to the same enzyme form as the variable substrate, the pattern will be competitive, or have only a slope effect in the double-reciprocal plot. For example in the ordered bireactant system the last product released inhibits competitively, when the first substrate to bind is the variable substrate.

2. If the product binds to an enzyme form that is upstream in a rapid-equilibrium segment from the enzyme form to which the variable substrate binds, the pattern will be competitive. For example in the bireactant, ordered, rapid-equilibrium chemical model the inhibition by the first product to be released was competitive when either the first or the second substrate to bind was the variable substrate.

3. The initial velocity, when the concentration of a given substrate is varied, will be inhibited noncompetitively, both slope and intercept effect, if the product binds only to a different enzyme form than that bound by the variable substrate, if the two forms are connected by reversible, steady-state steps. For example in the sequential, ordered, steady-state chemical model the last product to be released inhibits noncompetitively, when the second substrate to bind is the variable substrate, and the first product to be released also inhibits noncompetitively, when either the first or the second substrate is the variable substrate.

4. An uncompetitive pattern is predicted only when the enzyme form to which the product binds is isolated from that to which the variable substrate binds by irreversible steps both upstream and downstream. The irreversible step may be due to product release, in the absence of that product, or to an irreversible chemical (*e.g.* catalytic) step. The presence of one of the substrates at near-saturating concentration would ordinarily result in an irreversible step, but it should be realized that when the product inhibition is competitive with that substrate, the high concentration eliminates the inhibition altogether. For example in the sequential, ordered, steady-state chemical model inhibition by the first product to be released is noncompetitive, when the first substrate to bind is the variable substrate, but the inhibition becomes uncompetitive in the presence of a saturating concentration of the second substrate to bind. However, the ordinarily noncompetitive inhibition by the last product released, when the second substrate to bind is the variable substrate, is eliminated in the presence of saturating concentration of the first substrate to bind, because the inhibition is competitive with respect to the latter substrate.

In addition to all of the exceptions described above a number of additional factors listed at the beginning of this chapter can also affect the patterns of product inhibition.

7.5. Terreactant Models

Terreactant models are even more complex, mostly because of the greater number of possibilities. In order to substantiate the hypothesis of a particular chemical model by mathematical modeling the alternative models must be eliminated. With terreactant systems the derivation of all of the other possible mathematical models becomes an onerous process. Therefore, the investigation of product inhibition with these systems is most appropriate as a confirmation of a hypothesized chemical model that has been substantiated already by other methods.

In the interests of time and space the treatment in the present chapter will be limited to the application of the general principles above to a somewhat complex chemical model with verification by derivation.

7.5.1. TERREACTANT, SEQUENTIAL, ORDERED MODEL

A terreactant chemical model with steady-state, sequential, ordered substrate binding and product release (Table 7.3) will have nine inhibition patterns. Inhibition by the last product to release, R (Figure 7.11), is predicted to be competitive, when the first substrate to bind is the variable substrate; and noncompetitive, when either the second or the third substrate to bind is the variable substrate (equation 7.9).

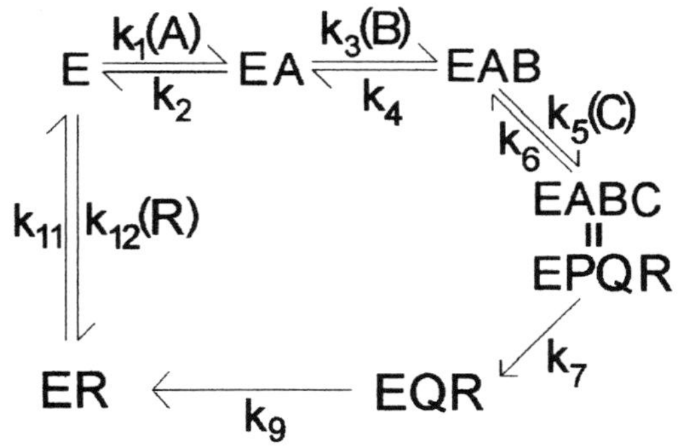

$$E \underset{k_2}{\overset{k_1(A)}{\rightleftharpoons}} EA \underset{k_4}{\overset{k_3(B)}{\rightleftharpoons}} EAB \underset{k_6}{\overset{k_5(C)}{\searrow}}$$

$$k_{11} \Big\| k_{12}(R)$$

$$EABC$$
$$\|$$
$$EPQR$$

$$ER \underset{k_9}{\longleftarrow} EQR \overset{k_7}{\swarrow}$$

Figure 7.11

$$\frac{v_i}{E_t} = \frac{1}{D}$$

$$D = \frac{1}{(A)} * \left[\frac{1}{k_1} + \frac{(R)}{K_{11}*k_1} \right] + \frac{1}{(A)*(B)} * \left[\frac{1}{K_1*k_3} + \frac{(R)}{K_{11}*K_1*k_3} \right] +$$

$$\frac{1}{(B)} * \frac{1}{k_3} + \frac{1}{(B)*(C)} * \left[\frac{1}{K_3*k_5} + \frac{1}{K_3*K_5*k_7} \right] + \frac{1}{(C)} * \left[\frac{1}{k_5} + \frac{1}{K_5*k_7} \right] + \tag{7.9}$$

$$\frac{1}{(A)*(B)*(C)} * \left[\frac{(R)}{K_{11}*K_1*K_3*k_5} + \frac{(R)}{K_{11}*K_1*K_3*K_5*k_7} + \frac{1}{K_1*K_3*k_5} + \frac{1}{K_1*K_3*K_5*k_7} \right] +$$

$$\frac{1}{k_7} + \frac{1}{k_9} + \frac{1}{k_{11}}$$

In the presence of the second product to be released the initial velocity is predicted to be inhibited uncompetitively, when any of the three substrates is the variable substrate, since this product is bounded both upstream and downstream by an irreversible step (Figure 7.12). The mathematical model verifies this prediction. (equation 7.10).

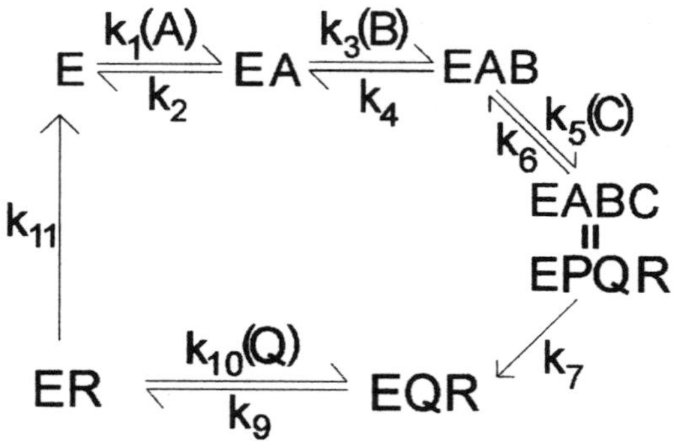

Figure 7.12

$$\frac{v_i}{E_t} = \frac{1}{D}$$

$$D = \frac{1}{(A)} * \frac{1}{k_1} + \frac{1}{(A)*(B)} * \frac{1}{K_1*k_3} + \frac{1}{(B)} * \frac{1}{k_3} + \frac{1}{(B)*(C)} * \left[\frac{1}{K_3*k_5} + \frac{1}{K_3*K_5*k_7} \right] +$$

$$\frac{1}{(C)} * \left[\frac{1}{k_5} + \frac{1}{K_5*k_7} \right] + \frac{1}{(A)*(B)*(C)} * \left[\frac{1}{K_1*K_3*k_5} + \frac{1}{K_1*K_3*K_5*k_7} \right] +$$

$$\frac{1}{k_7} + \frac{1}{k_9} + \frac{(Q)}{K_9*k_{11}} + \frac{1}{k_{11}}$$

(7.10)

Inhibition by the first product released is predicted to be noncompetitive, when either of the substrates is the variable substrate, since it is a downstream product inhibitor connected by reversible steps (Figure 7.13) and this prediction is verified by the mathematical model (equation 7.11).

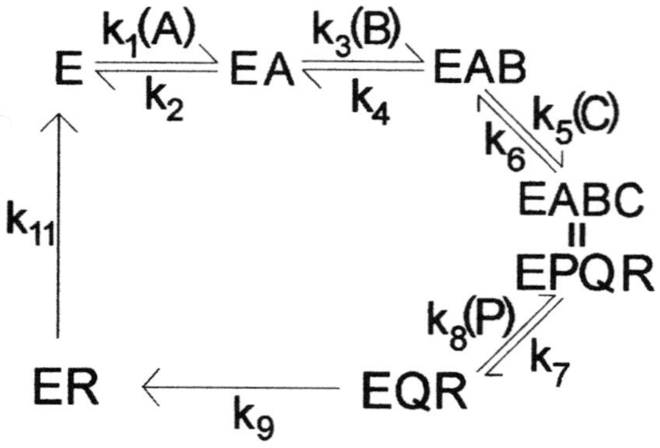

Figure 7.13

$$\frac{v_i}{E_t} = \frac{1}{D}$$

$$D = \frac{1}{(A)} * \frac{1}{k_1} + \frac{1}{(A)*(B)} * \frac{1}{K_1 * k_3} + \frac{1}{(B)} * \frac{1}{k_3} +$$

$$\frac{1}{(B)*(C)} * \left[\frac{1}{K_3 * k_5} + \frac{1}{K_3 * K_5 * k_7} + \frac{(P)}{K_3 * K_5 * K_7 * k_9} \right] + \frac{1}{(C)} * \left[\frac{1}{k_5} + \frac{1}{K_5 * k_7} + \frac{(P)}{K_5 * K_7 * k_9} \right] +$$

$$\frac{1}{A)*(B)*(C)} * \left[\frac{1}{K_1 * K_3 * k_5} + \frac{1}{K_1 * K_3 * K_5 * k_7} + \frac{(P)}{K_1 * K_3 * K_5 * K_7 * k_9} \right] + \frac{1}{k_7} + \frac{(P)}{K_7 * k_9} + \frac{1}{k_9} + \frac{1}{k}$$

(7.11)

Table 7.3: Initial velocity patterns with product inhibitors in ternary, sequential, ordered enzyme reactions. C=competitive, UC= uncompetitive, NC=noncompetitive.

Variable Substrate	Product		
	P	Q	R
A	NC	UC	C
B	NC	UC	NC
C	NC	UC	NC

In Chapter 5 evidence was summarized for ordered binding of MgATP (A), bicarbonate (B) and ammonia (C) to the enzyme carbamoyl-phosphate synthetase [6] (Figure 7.14). Although bicarbonate was the second substrate to bind, the order of the other two substrates could be determined only with additional experiments. The steady-state mechanism is complicated somewhat by the fact that two molecules of

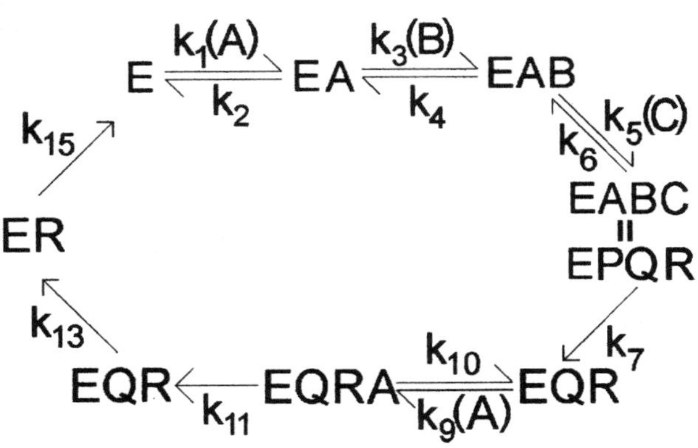

Figure 7.14

MgATP are utilized with somewhat different fates of the γ-phosphate, although both form MgADP. Furthermore, glutamine can be an alternative source of ammonia in the reaction. The results that product inhibition by phosphate is competitive, when the concentration of glutamine is 0.1 mM; but becomes uncompetitive, when the concentration is raised fifty fold show that glutamine (ammonia) binds in a step between the binding of bicarbonate and the release of phosphate. Therefore, the supported hypothesis for substrate binding is: MgATP, bicarbonate, glutamine (ammonia); followed by the release of inorganic phosphate. In addition the result that the inhibition by inorganic phosphate (Figure 7.14) is competitive, when MgATP is the variable substrate indicates that the both bind to the same enzyme form (*c.a.* equation 7.12). Therefore, the supported hypothesis is supplemented with the binding of an additional MgATP immediately after the release of inorganic phosphate. Furthermore, the fact that the inhibition by carbamoyl phosphate is uncompetitive, when either MgATP, bicarbonate or ammonia is the variable substrate indicates that its release step is bounded on both sides by steps in which other products are released, namely two molecules of MgADP. Therefore the supported hypothesis is that in Figure 7.14. Although the data are consistent with the hypothesized chemical model, the process to eliminate of all of the other possible chemical models becomes somewhat daunting.

7.6. Summary

Investigations of product inhibition are technically straight forward, although somewhat laborious, if done completely. They provides an approach to the study of product release. However, exceptions to the basic inhibition patterns can be produced by unrecognized irreversible steps, the formation of abortive complexes, and the existence of enzyme changes (isomerizations) associated with product release. The multiplicity of possible chemical models due to these possible exceptions as well as those due to the possible models for product release makes the strict elimination of all but one chemical model a rather daunting prospect, that is generally accomplished only for bisubstrate:biproduct reactions that seem to have no exceptional patterns. For more complex systems it provides confirmatory evidence for hypotheses that have already good evidence, *e.g.* analog inhibition.

7.7. References

1. Rudolph, F.B. "Product Inhibition and Abortive Complex Formation," *Methods Enzymol,*63, 411-36 (1979).

2.Munagala, N.R., Chin, M.S. and Wang, C.C. "Steady-State Kinetics of the Hypoxanthine-Guanine-Xanthine Phosphoribosyltransferase from *Tritrichomonas foetus*: The Role of Threonine-47," *Biochemistry*, 37, 4045-51, (1998).

3.Sanghani, P.C., Stone, C.L., Ray, B.D., Pindel, E.V. Hurley, T.D. and Bosron, W.F. "Kinetic Mechanism of Human Glutathione-Dependent Formaldehyde Dehydrogenase," *Biochemistry*, 39, 10720-9. (2000).

4. Cleland, W.W. "Determining the Chemical Mechanisms of Enzyme-Catalyzed Reactions by Kinetic Studies," *Adv. Enzymol.* 45, 288-92 (1977).

5.Wong, L.-J. And Frey, P.A. "Galactose-1-phosphate Uridylyltransferase: Rate Studies Confirming a Uridylyl-Enzyme Intermediate on the Catalytic Pathway," *Biochemistry*, 13, 3889-94 (1974).

6. Raushel, F.M., Anderson, P.M. and Villafranca, J.J. "Kinetic Mechanism of *Escherichia coli* Carbamoyl-Phosphate Synthetase," *Biochemistry*, 5587-91 (1978).

CHAPTER 8

EFFECTS OF SUBSTRATE INHIBITION

8.1. Introduction

One of the most interesting and useful phenomena in steady-state enzyme kinetics is substrate inhibition and the complementary phenomenon, substrate activation. Although it first appears to be an annoyance in experiments in which substrate concentration is varied, systematic investigation will identify minor pathways and nonproductive enzyme complexes. Investigators who avoid using data at the inhibiting levels of substrate concentration are avoiding data that can provide evidence for nonproductive complexes and for minor pathways in the reaction at physiologic concentrations of substrate.

Substrate inhibition and activation are different from negative and positive cooperativity respectively, homotropic effects, which require multiple, interacting active sites. Cooperativity is a normal physiological phenomenon, and is not a subject of this book. Substrate inhibition and activation are commonly demonstrated with nonphysiologic concentrations of substrate and can be seen with monomeric enzymes. Nevertheless it is interesting to speculate that some of the principles presented below might be useful in the elucidation of detailed mechanisms of homotropic effects.

8.2. Experiments and Results

Substrate inhibition is seen in the results of experiments in which the initial velocity is measured as a function of substrate concentration. At high substrate concentration the velocity decreases with increasing substrate concentration or at least the velocity does not increase as much as it should. The investigator should be sensitive to the possibility that high levels of the substrate may inhibit the rate for reasons related more directly to things such as pH and ionic strength than to binding of substrate *per se*, particularly if

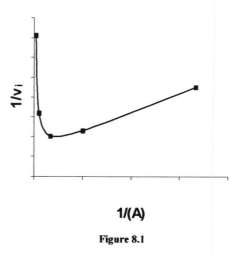

Figure 8.1

the inhibiting concentration of substrate is above 100 millimolar or if the inhibition is rather weak. In any case it is a good idea to measure the pH of the reaction mixture in the presence of inhibiting concentrations of substrate. The effects of ionic strength can be tested by the addition of salt to the reaction mixture so that the ionic strength is comparable.

Graphically substrate inhibition is most easily seen in a double-reciprocal plot of the data, where the values for $1/v_i$ increase close to the vertical axis (Figure 8.1). The points seem to describe a line that "hooks up" close to the vertical axis. Frequently the remainder of the points, the non-inhibiting, points approximately describe a straight line that can be regarded as a noninhibited set of data for the purposes of other substrate studies.

The inhibition may be complete or incomplete. Complete inhibition generally indicates the formation of an nonproductive complex, whereas incomplete inhibition generally indicates the existence of a minor pathway. For the purpose of the present discussion the complexes involved in both will be referred to as less-productive with the realization that the adjective also includes nonproductive. The double-reciprocal plot of a data set describing complete inhibition will approach the vertical axis as an asymptote, whereas that describing incomplete inhibition will cross the axis at some point. However, the completeness of the inhibition is best demonstrated by a plot of the reciprocal of the initial velocity, $1/v_i$, *vs* the substrate concentration in the inhibiting range. The demonstration that the points tend to become an ascending straight line at high concentration indicates complete inhibition, whereas the demonstration that the points tend to become nonlinear and approach a horizontal line indicates incomplete inhibition.

8.3 Interpretation and Models

Substrate inhibition is most easily interpreted in the context of previous knowledge about the order of binding and release of substrates and products. Whereas substrate inhibition is generally caused by the formation of a non-productive, or less productive, complex between some enzyme form and the substrate, substrate activation is generally caused by the formation of a more productive complex between some enzyme form and the substrate. Further investigations of both substrate inhibition and substrate activation are focused on the identification of the enzyme form to which the substrate binds to form the complex in question.

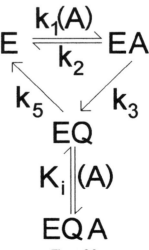

Figure 8.2

8.3.1. UNIREACTANT MODELS

Although substrate inhibition in a unireactant system is generally less frequently investigated than multisubstrate enzymes, there are a couple of interesting chemical models to be distinguished.

In chemical models with one substrate and two products, *e.g.* lyases, hydrolases, the substrate may bind to the enzyme-product complex (Figure 8.2). The corresponding mathematical model (equation 8.1) can be derived by a similar algorithm to that used for derivation of models with analog inhibitors.

$$\frac{v_i}{E_t} = \frac{1}{\frac{1}{(A)}*\left[\frac{1}{k_1}*\frac{1}{K_1*k_3}\right]+\frac{1}{k_3}+\frac{1}{k_5}*\left[1+\frac{(A)}{K_I}\right]}$$

$$v_i = \frac{(A)}{\frac{K_A}{V_{max}}*\frac{(A)^2}{K_{Iapp}}+\frac{(A)}{V_{max}}}$$

(8.1)

Alternatively, there are enzymes to which two molecules of substrate bind to the active site to form a less-productive complex (Figure 8.3). However, the mathematical models (compare equation 8.2) derived from each of these chemical models have the same

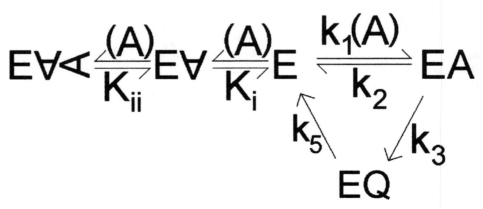

Figure 8.3

overall form as equation 8.1. Therefore, they cannot be distinguished with data from experiments in which only the substrate concentration is varied.

$$\frac{v_i}{E_t} = \frac{1}{\frac{1}{(A)} * \left[\frac{1}{K_1} + \frac{1}{K_1 * k_3}\right] * \left[1 + \frac{(A)}{K_i} + \frac{(A)^2}{K_i * K_{ii}}\right] + \frac{1}{k_3} + \frac{1}{k_5}}$$

$$v_i = \frac{1}{\frac{1}{(A)} * \frac{K_A}{V_{max}} + \frac{K_A}{V_{max} * K_i} + \frac{(A) * K_A}{V_{max} * K_i * K_{ii}} + \frac{1}{V_{max}}} \qquad (8.2)$$

$$v_i = \frac{(A)}{\frac{K_A}{V_{max}} + \frac{(A)^2 * K_A}{V_{max} * K_i * K_{ii}} + \frac{K_A * (A)}{V_{max} * K_i} + \frac{(A)}{V_{max}}}$$

However, Cleland [1] has described product inhibition experiments that can distinguish the two. In any case, the equation (Equation 8.1) is useful for curve fitting purposes.

8.3.2. BIREACTANT MODELS

Investigations of substrate inhibition (*e.g.* Figure 8.4) are more complicated and interesting in bireactant enzyme systems than in unireactant ones. With bireactant and other multireactant models it is most useful to focus on data of the initial velocity with the

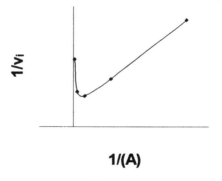

Figure 8.4

noninhibiting substrate as the variable substrate and the inhibiting substrate as the constant-variable substrate (*e.g.* Figure 8.5). Although the resulting double-reciprocal plots appear, initially, quite confusing, the replot of the slope, reciprocal V_{max}/K_M, and intercept, reciprocal V_{max}, respectively *vs* the reciprocal of the concentration of the inhibiting substrate will determine whether the inhibition affects the slope, the intercept, or both; *i.e.* whether the inhibition is competitive, uncompetitive or

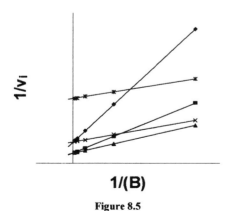

1/(B)

Figure 8.5

noncompetitive with respect to the inhibiting substrate (*e.g.* Figure 8.6).

   In practice it is best to fit the data set corresponding to each different concentration of the inhibiting, constant-variable, substrate to the Michaelis-Menten equation to determine the best value for the slope and intercept of each. It is almost always sufficient to plot these latter values *vs* the reciprocal of the concentration of the inhibiting substrate to determine which set of values shows the inhibition.   In equivocal cases the reciprocal of the parameter values (V_{max} or V_{max}/K_M) can be fitted to the Michaelis-Menten equation and to the substrate-inhibition equation (equation 8.1) to see which one fits best in each case.

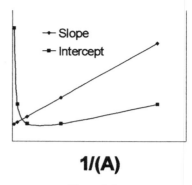

1/(A)

Figure 8.6

Bireactant, Sequential, Ordered Model
In a bisubstrate, sequential, ordered system the most common cause of substrate inhibition is the formation of a less-productive complex between the substrate and the binary enzyme-product complex.   Most frequently the second substrate to bind forms a less-productive complex with the binary enzyme-product complex (Figure 8.7), the corresponding mathematical model (equation 8.3) predicts that the inhibition will be demonstrated in the intercept but not the slope of the double-reciprocal plot, when the

$$E \underset{k_2}{\overset{k_1(A)}{\rightleftharpoons}} E$$

$$k_7 \qquad k_4 \| k_3(B)$$

$$EAB$$
$$\|$$
$$EQB \underset{(B)}{\overset{K_i}{\rightleftharpoons}} EQ \underset{k_5}{\longleftarrow} EPQ$$

Figure 8.7

noninhibiting substrate is the variable substrate and the inhibiting substrate is the constant-variable substrate. Although the model above was constructed with the second substrate as the inhibiting substrate, the same results are achieved, when the first substrate is inhibiting. Other, less common, models for substrate inhibition can be derived in similar manner. The inhibition patterns in the model above (Figure 8.7) are understandable when one realizes that this is consistent with one of the rules for analog inhibitors stated in Chapter 6 that only an intercept effects results, when the enzyme form to which the

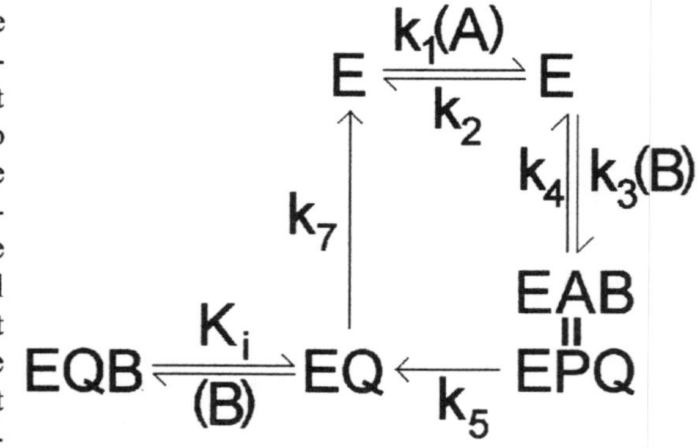

$$\frac{v_i}{E_t} = \frac{1}{D}$$

$$D = \frac{1}{(A)} * \frac{1}{k_1} + \frac{1}{(A)*(B)} * \left[\frac{1}{K_1 * k_3} + \frac{1}{K_1 * K_3 * k_5}\right] + \frac{1}{(B)} * \left[\frac{1}{K_3} + \frac{1}{K_3 * k_5}\right] + \qquad (8.3)$$

$$\frac{1}{k_5} + \frac{1}{k_7} * \left[1 + \frac{(B)}{K_i}\right]$$

substrate binds as an inhibitor is downstream and separated by irreversible steps, *i.e.* product release, from the form to which either substrate binds as a substrate.

For example the secondary-alcohol dehydrogenase from the thermophilic organism, *Thermoanaerobium brockii*, was shown in Chapter 6 to bind NADPH and cyclopentanone in that order [2]. The initial velocity is inhibited at high concentrations of the substrate, cyclopentanone at 25 °C. The results of initial-velocity experiments with the noninhibiting substrate, NADPH, as the variable substrate and cyclopentanone as the constant-variable substrate demonstrated that the intercept but not the slope of the double-reciprocal plot showed the inhibition. These results provided evidence that cyclopentanone was forming a complex with the enzyme-NAD product complex, for which additional evidence is described below.

Other mathematical models for substrate inhibition of bisubstrate, sequential, ordered chemical models can be derived by similar means, but none predict this particular pattern of inhibition. For example the model in which a second molecule of the first substrate forms a less-productive complex with the binary complex of itself as a substrate and the enzyme (*e.g.* EAA) results in the prediction of only a slope effect, when the second substrate, B, is the variable substrate and A is the constant-variable substrate.

Induced Substrate Inhibition. Induced substrate inhibition is seen with enzymes that conform to a sequential, ordered chemical model. It is the demonstration of substrate inhibition only in the presence of an analog-inhibitor to one of the substrates. Its demonstration is confirmation of order in the binding of substrate. The underlying chemical model (Figure 8.8) includes an analog

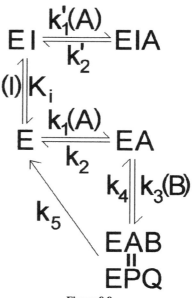

Figure 8.8

$$\frac{v_i}{E_t} = \frac{1}{D}$$

$$D = \frac{1}{(A)} * \frac{1}{k_1} * \left[1 + \frac{(I)}{K_I} + \frac{(I)*(A)}{K_I*K_{SI}}\right] + \tag{8.4}$$

$$\frac{1}{(A)*(B)} * \left[\frac{1}{K_1*k_3} + \frac{1}{K_1*K_3*k_5}\right] * \left[1 + \frac{(I)}{K_I} + \frac{(I)*(A)}{K_I*K_{SI}}\right] + \frac{1}{(B)} * \left[\frac{1}{K_3} + \frac{1}{K_3*k_5}\right] + \frac{1}{k_5}$$

inhibitor to the first substrate to bind and the binding of the substrate to the complex of inhibitor with free enzyme. The corresponding mathematical model (equation 8.4) supports the prediction of substrate inhibition by the first substrate to bind and that inhibition will be manifest in the slope but not the intercept (competitive), when the noninhibiting substrate, B, is the variable substrate and the inhibiting substrate, A, is the constant-variable substrate.

Thus, the analysis of this substrate inhibition not only confirms the hypothesis of an ordered model but also the sequence of that order.

The enzyme phosphoenolpyruvate carboxylase, from *Zea mays* catalyzes the substitution of bicarbonate into phosphoenolpyruvate to yield oxaloacetate and inorganic phosphate. Phosphoglycolate, and analog of phosphoenolpyruvate induces substrate inhibition by bicarbonate [3]. This inhibition confirms the hypothesis that the enzyme mechanism conforms to an ordered model with phosphoenolpyruvate and bicarbonate adding in that order. When phosphoenol pyruvate is the variable substrate and bicarbonate is the constant-variable substrate in the presence of a constant concentration of phosphoglycolate, only the slope of the double-reciprocal plot is affected. Thus, the inhibition is competitive with respect to phosphoenolpyruvate. In addition the hyperbolic nature of the slope effect indicates that the inhibition is partial, which, in turn, implies the existence of a minor pathway with the dissociation of the inhibitor and addition of the phosphoenolpyruvate as the second substrate instead of the first.

Bireactant, Sequential, Random Models
A chemical model of bisubstrate, sequential, rapid-equilibrium, random substrate addition and product release should not include complexes of substrate with binary enzyme-product complexes, since the concentration of the latter should be vanishingly small. However it must be remembered that the rapid-equilibrium condition is sometimes more of an approximation than a reality. The substrate inhibition in such a model should affect the intercept but not the slope of a double-reciprocal plot, when the noninhibiting substrate is the variable substrate.

Furthermore, the rapid-equilibrium models may show substrate inhibition due to the formation of a less-productive complex of the inhibiting substrates with the binary complex of itself as substrate with the free enzyme (Figure 8.9). Thus two molecules of the inhibiting substrate have bound to the enzyme usually with the second molecule binding to the site ordinarily and productively occupied by the noninhibiting substrate. This chemical model results in a mathematical model (equation 8.5) that supports the prediction of an inhibition pattern in which the slope but not the

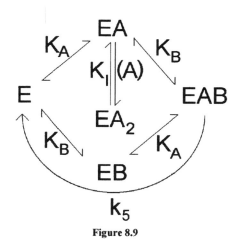

Figure 8.9

$$\frac{v_i}{E_t} = \frac{1}{\frac{1}{k_5} * \left[1 + \frac{K_A}{(A)} + \frac{K_B}{(B)} + \frac{K_A * K_B}{(A) * (B)} + \frac{K_B * (A)}{(B) * K_I} \right]} \quad (8.5)$$

intercept of the double-reciprocal plot will be affected, when the noninhibiting substrate is the variable substrate and the inhibiting substrate is the constant-variable substrate, *i.e.* competitive substrate inhibition.

Bireactant, Nonsequential Models
Substrate inhibition is very common in investigations of single-site, bisubstrate, double-displacement systems. It is intuitively somewhat obvious that if both substrates bind to the same site a molecule of one of them might bind to that site, after another molecule of the same substrate had already changed the enzyme as long as the change is not too great, *e.g.* added a small chemical group to the enzyme, in preparation for the binding of the other substrate. In fact if substrate inhibition is not demonstrated in investigation of an enzyme hypothesized to conform to a double-displacement model, it is wise to reexamine the previous evidence for the double-displacement model. In addition the analysis is sometimes complicated by double substrate inhibition, inhibition by both substrates.

Substrate inhibition is less common with enzymes that conform to a double-displacement model but have different active sites for each substrate. Therefore, the absence of substrate inhibition in the face of other convincing evidence for a double-displacement model would constitute preliminary, circumstantial evidence for two different active sites.

The most common chemical model (Figure 8.10) results in a mathematical model (equation 8.6) that supports the prediction that the inhibition will be manifest in the slope but not the intercept (competitive inhibition) of the double-reciprocal plot, when the noninhibiting substrate is the variable substrate and the inhibiting substrate is the constant-variable substrate.

The enzyme O-acetylserine sulfhydrylase from *Salmonella typhimurium* catalyzes the substitution of sulfide for the O-acetyl group to yield cysteine. Analog inhibitor evidence was presented in Chapter 6 that the enzyme conforms to a double displacement model [4]. That hypothesis is confirmed by the demonstration of competitive substrate inhibition by each of the substrates, sulfide and O-acetylserine.

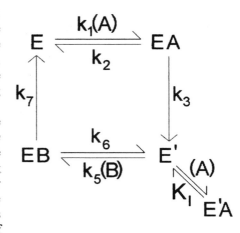

Figure 8.10

$$\frac{v_i}{E_t} = \frac{1}{\frac{1}{(A)}*\left[\frac{1}{k_1}+\frac{1}{K_1*k_3}\right]+\frac{1}{(B)}*\left[\frac{1}{k_3}+\frac{1}{K_3*k_5}\right]*\left[1+\frac{(A)}{K_I}\right]+\frac{1}{k_3}+\frac{1}{k_5}} \qquad (8.6)$$

Summary of Bisubstrate Models
It can be seen from the previous examples that the rules for substrate inhibition are very similar to those for analog inhibitors with the additional conditions that the inhibitor is also the substrate and that the noninhibiting substrates are the variable substrates.

1. Thus the inhibition will be competitive, slope effect only, if the inhibiting substrate forms a less productive complex with the same enzyme form to which the variable substrate binds.

2. The inhibition will be uncompetitive, intercept effect only, if the inhibiting substrate forms a less productive complex with an enzyme form downstream from the variable substrate or if that enzyme form is connected with the enzyme form to which the variable substrate binds by irreversible steps.

3. The inhibition will be noncompetitive, both slope and intercept effects, if the inhibiting substrate forms a less productive complex with an enzyme form upstream from the variable substrate and that is converted to the enzyme form to which the variable substrate binds by reversible steps.

8.3.3. PRODUCT INHIBITION REVISITED

Evidence for the identity of less-productive, including nonproductive, enzyme-substrate complexes can also result from product inhibition experiments, which will complement the evidence from substrate inhibition experiments. The product inhibition patterns are determined for the reaction in the opposite direction, and the determination is made whether the slope (*i.e.* reciprocal V_{max}/K_M), the intercept *(i.e.* reciprocal V_{max}), or both the slope and the intercept of the double-reciprocal plot are linear or parabolic *vs* the concentration of product inhibitor. The fact that the inhibition patterns with less-productive complexes will be somewhat different from those described in the previous chapter reinforces the necessity to determine all of the product inhibition patterns and to determine whether the slope, the intrcept, or both, of the double reciprocal plot of each pattern is linear or parabolic.

In practice the determination is best accomplished by fitting each data set collected with a single concentration of product and several concentrations of the variable substrate to the Michaelis-Menten equation to determine the value of the V_{max}/K_M and of the V_{max} of each.. Then fit each series of values of the reciprocal of each of these *vs* the product concentration to both a first and second order polynomial equation to determine whether either or both is linear or parabolic. Alternatively the operations can be accomplished graphically, although the determinations are not as rigorous.

The chemical models are too numerous and the interpretation too complicated to present all of the possibilities here. Therefore, the discussion will be limited to some of the more common models to demonstrate the approach and to illustrate the range of possibilities. However, derivation of the mathematical models from the chemical models is accomplished by the same algorithms used previously.

Bireactant, Biproduct, Sequential, Ordered Model
The reverse reaction of substrate inhibition with an enzyme that conforms to a bisubstrate, sequential, ordered model, in which the second substrate to bind also binds to the binary complex of the enzyme and the last product released (Figure 8.7) is a model of product-inhibition patterns for a bisubstrate, sequential, completely ordered chemical model with the formation of a less-productive complex due the combination of the first product released with the complex of the enzyme with the first substrate to bind (Figure

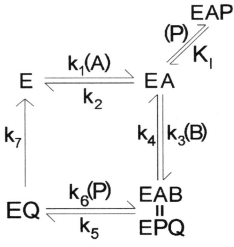

Figure 8.11

8.11). This model results in a mathematical model (equation 8.7) that supports the prediction of noncompetitive inhibition, when the second substrate, B, is the variable substrate in the presence of the first product released. The presence in the slope term for

$$\frac{v_i}{E_t} = \frac{1}{D}$$

$$D = \frac{1}{(A)} * \frac{1}{k_1} + \frac{1}{(A)*(B)} * \left[\frac{1}{K_1 * k_3} + \frac{1}{K_1 * K_3 * k_5} + \frac{(P)}{K_1 * K_3 * K_5 * k_7} \right] +$$

$$\frac{1}{(B)} * \left[\frac{1}{k_3} + \frac{1}{K_3 * k_5} + \frac{(P)}{K_3 * K_5 * k_7} \right] * \left[1 + \frac{(P)}{K_I} \right] + \frac{1}{k_5} + \frac{1}{k_7}$$

(8.7)

the double-reciprocal plot of product concentration times itself (second order) shows that the replot of the slope will be parabolic. The same mathematical model supports the prediction of noncompetitive inhibition with a parabolic intercept, when the first substrate, A, is the variable substrate in the presence of the first product released.

For example as a part of the investigation of the substrate inhibition of secondary-alcohol dehydrogenase from *Thermoanaerobium brockii* described above the product inhibition patterns [2] by cyclopentanone were determined to be noncompetitive with linear intercept and parabolic slope effects, when either cyclopentanol or NADP is the variable substrate. Thus these results conform to the chemical model described above.

However, the formation of less-productive complexes with product does not always result in parabolic inhibition. In some models it merely changes the pattern of linear inhibition. For example the product-inhibition patterns for a bisubstrate sequential, completely ordered chemical model with the formation of a less-productive complex due to the combination of the last product released with the complex of the enzyme and the first substrate to bind (Figure 8.12) results in a mathematical model (equation 8.8) that establishes the prediction of noncompetitive linear patterns for both substrates. Although they are linear, the latter patterns are different from those in the absence of the less-productive complex, which were

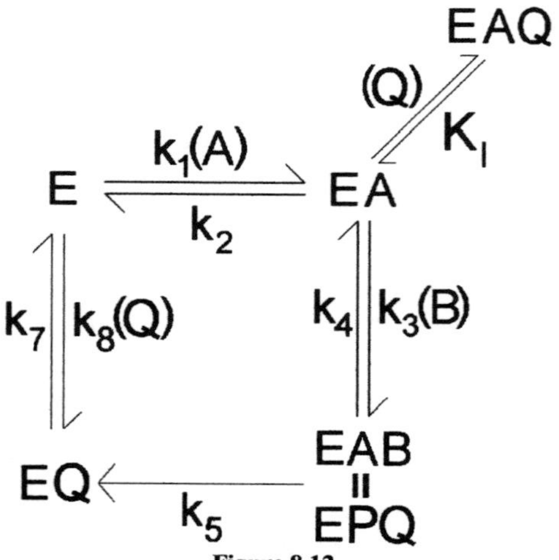

Figure 8.12

$$\frac{v_i}{E_t} = \frac{1}{D}$$

$$D = \frac{1}{(A)}*\left[\frac{1}{k_1}+\frac{(Q)}{K_7*k_1}\right]+\frac{1}{(A)*(B)}*\left[\frac{1}{K_1*k_3}+\frac{(Q)}{K_7*K_1*k_3}+\frac{1}{K_1*K_3*k_5}+\frac{(Q)}{K_7*K_1*K_3*k_5}\right]+ \quad (8.8)$$

$$\frac{1}{(B)}*\left[\frac{1}{k_3}+\frac{1}{K_3*k_5}\right]*\left[1+\frac{(Q)}{K_I}\right]+\frac{1}{k_5}+\frac{1}{k_7}$$

competitive and noncompetitive for the first and second substrates respectively. Therefore, if the order of binding of substrate were established in both directions, the product inhibition data would provide evidence for the less-productive complex.

Bireactant, Sequential, Random Models

In theory, bireactant chemical models with rapid-equilibrium, random binding of substrates and products do not show substrate inhibition due to the formation of adducts of substrate with the binary enzyme-product complexes. However, in reality product can form less-productive complexes with enzyme-substrate forms, in product-inhibition experiments. The prediction of product-inhibition patterns in the latter case (Figure 8.13), can be derived

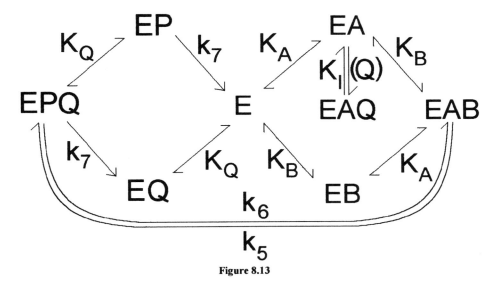

Figure 8.13

in similar manner to those with the ordered models above. Thus, the corresponding mathematical model (equation 8.9) supports the prediction of a noncompetitive pattern, when the first substrate to bind (A) is the variable substrate and a competitive pattern, when the second substrate to bind (B) is the variable substrate.

$$\frac{v_i}{E_t} = \frac{1}{\frac{1}{k_5}*\left[1+\frac{K_A}{(A)}+\frac{K_B}{(B)}*\left[1+\frac{(Q)}{K_I}\right]+\frac{K_A*K_B}{(A)*(B)}*\left[1+\frac{(Q)}{K_Q}\right]\right]} \tag{8.9}$$

Thus, although product inhibition in a rapid-equilibrium, random chemical model without a less-productive complex results in four competitive patterns, the presence of such a complex converts these patterns to three competitive patterns and one noncompetitive pattern.

Since the random chemical model is symmetric, the formation of a less-productive complex with the other enzyme-substrate complex would result in a noncompetitive pattern, when that substrate is the variable substrate. Thus the identity of both the product and the substrate components of the less-productive complex can be established from analysis of the data from product-inhibition experiments.

For example evidence was presented in Chapter 6 that the enzyme 6-phosphogluconate dehydrogenase from *Candida utilis* has a mechanism that conforms to a random, rapid-equilibrium model [5]. The same investigation demonstrated that product inhibition between similar substrate-product pairs, NADP/NADPH and ribulose-5-phosphate/6-phosphogluconate are competitive, whereas those between dissimilar substrate-product pairs, NADP/ribulose-5-phosphate and NADPH/6-phosphogluconate, are noncompetitive. The latter patterns constitute evidence that both dissimilar substrate-product pairs form less-productive complexes with the enzyme.

Bireactant, Nonsequential Model
The prediction of product-inhibition patterns in bireactant, double displacement chemical models with the formation of an enzyme-product less-productive complex (Figure 8.14), can also be derived in similar manner. The corresponding mathematical model (equation 8.10) supports the prediction of noncompetitive inhibition with a parabolic slope effect, when either the first or the second substrate is the variable substrate. This model is common with enzymes that conform to a double-displacement model and have only a single kind of active site. These

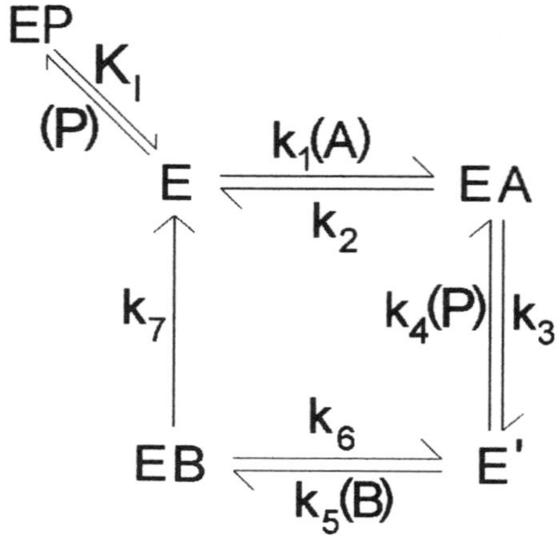

Figure 8.14

patterns are in contrast to one competitive and one noncompetitive pattern, when the first and second substrates respectively are the variable substrates, in the absence of a less-productive complex.

$$\frac{v_i}{E_t} = \frac{1}{D}$$

$$D = \frac{1}{(A)} * \left[\frac{1}{k_1} + \frac{1}{K_1 * k_3}\right] * \left[1 + \frac{(P)}{K_I}\right] + \frac{1}{(A)*(B)} * \left[\frac{(P)}{K_1 * K_3 * k_5} + \frac{(P)}{K_1 * K_3 * K_5 * k_7}\right] * \left[1 + \frac{(P)}{K_I}\right] + \text{(8.10)}$$

$$\frac{1}{(B)} * \left[\frac{1}{k_5} + \frac{1}{K_5 * k_7} + \frac{(P)}{K_3 * k_5} + \frac{(P)}{K_3 * K_5 * k_7}\right] + \frac{1}{k_3} + \frac{1}{k_7}$$

It is tempting to set down rules for the prediction of the inhibition patterns due to product inhibition with the formation of less-productive complexes of some enzyme form with product, but exceptions abound. Therefore, it seems best to derive the initial-velocity expression for each model under consideration. In a more global context it is best to have good evidence for the chemical model without the data from product-inhibition experiments. Then if the data from the latter experiments conforms to the chemical model without less-productive complexes, it confirms the original model. If not, the patterns associated with the formation of various less-productive complexes can be explored.

8.3.4. TERREACTANT MODELS

The experiments and the methods for data analysis and interpretation of data from terreactant enzyme systems are similar to those for bireactant systems. Here it will be sufficient to present the results and the interpretation of a specific enzyme.

Octopine dehydrogenase catalyses the secondary reductive amination of pyruvate by the α-amino group of arginine at the expense of NADH. Evidence was described in a previous chapter supporting the hypothesis that NADH binds to the enzyme first followed by pyruvate and arginine in random sequence. Complete substrate inhibition by both pyruvate and arginine was demonstrated [6]. That the inhibition by either affected only the intercepts, uncompetitive, when the other substrate was varied at noninhibitory levels indicates that pyruvate and/or arginine forms an inhibitory complex with an enzyme form downstream from the enzyme-NADH complex, probably to an enzyme-product complex.

Product inhibition by pyruvate of the reverse reaction, of octopine and NAD, was noncompetitive, when octopine was the variable substrate. The slope was parabolic but the intercept was linear with the concentration of pyruvate. These product-inhibition experiments indicate that the pyruvate forms a less-productive complex with the same enzyme form to which octopine binds, the enzyme-NAD complex. The demonstrated synergy of the substrate inhibition between pyruvate and arginine indicates that an adduct of the two is the inhibiting agent, *e.g.* the imine.

8.4. Summary

The investigator should be sensitive to the possibility of substrate inhibition and investigate it, when it appears. Substrate activation can also be interpreted in similar manner.

Determination whether the inhibition is complete or incomplete will provide evidence whether the inhibition is due to the formation of a nonproductive or merely a less-productive complex of some enzyme form with the substrate. Further experiments will probably clarify the identity of the other components of the complex and how it is assembled. Substrate activation can be interpreted to be due to the formation by the substrate and some enzyme form, of a complex that undergoes some rate-determining step (*e.g.* product dissociation) at a greater rate than the normal complex. The same kinds of experiments can clarify the identity of the complex.

Determination of the inhibition patterns with respect to the noninhibiting substrate will frequently help to identify whether the substrate binds as an inhibitor to the same enzyme form as does the noninhibiting substrate, upstream to it, or downstream from it.

Determination of the product-inhibition patterns of the reaction in the opposite direction will help further to identify the less-productive complex. Deviations of these patterns from those described in the previous chapter will further confirm the identity of the less-productive complex.

8.5. References

1. Cleland, W.W. "Substrate Inhibition," *Methods Enzymol.* 63, 500-513 (1979).

2. Ford, J.B., Askins, K.J. and Taylor, K.B. "Kinetic Models for Synthesis by a Thermophilic Alcohol Dehydrogenase," *Biotechnol. Bioeng.* 42, 367-75 (1993).

3. Jane, J.W., O'Leary, M.H. and Cleland, W.W. "A Kinetic Investigation of Phosphoenolpyruvate Carboxylase from *Zea mays. Biochemistry*, 31, 6421-6 (1992).

4. Tai, C.-H., Nalabolu, S.R., Jacobson, T.M., Minter, D.E. and Cook, P.F. "Kinetic Mechanisms of the A and B Isozymes of O-Acetylserine Sulfhydrylase from *Salmonella typhimurium* LT-2 Using the Natural and Alternative Reactants," *Biochemistry*, 32, 6433-42 (1993).

5. Berdis, A.J. and Cook, P.F. "Overall Kineic Mechanism of 6-Phosphogluconate Dehydrogenase from *Candida utilis*," *Biochemistry*, 32, 2036-40 (1993).

6. Schrimsher, J.L. and Taylor, K.B. "Octopine Dehydrogenae from *Pecten maximus*: Steady-State Mechanism," *Biochemistry*, 23, 1348-53 (1984).

CHAPTER 9

SLOW AND TIGHT INHIBITION

9.1. Introduction

A classification of inhibitors was presented in Chapter 6 the chapter which also contained a detailed discussion of the effects of rapid, reversible, analog inhibitors. That discussion of rapid, reversible, analog inhibitors required two assumptions and approximations that are not valid with other types of inhibitors. First the dissociation constant of the inhibitor was assumed to be sufficiently high that a wide variation of degrees of inhibition can be observed with the inhibitor concentration much greater than that of enzyme. Thus the unbound inhibitor concentration was assumed to be approximately that added to the reaction mixture. Second, The assumption was made that the inhibitor binds to the enzyme and inhibits it rapidly compared to the time during which the steady state approximations are valid. Thus the initial velocity could be measured immediately after the combination of enzyme and inhibitor.

The present chapter is a discussion of inhibitors for which one or both of these assumptions is invalid. Since this chapter is somewhat parenthetical to the remainder of the book, it can be skipped without prejudice to the understanding of the remaining issues. However, many of these inhibitors are physiological molecules that function in a number of control functions at the intracellular, cellular and tissue levels. Therefore, the chapter contains a discussion of one of the currently most important and exciting approaches to enzyme behavior.

The tight-binding inhibitors bind the enzyme so tightly, although reversibly, that the concentration of inhibitor necessary to observe intermediate degrees of inhibition (*e.g.* 50% inhibition) are comparable to the concentration of enzyme necessary to observe a measurable reaction rate. Therefore, a significant fraction of the total inhibitor is bound to the enzyme and the concentration of free inhibitor is not the same as the concentration of total inhibitor. Thus the development of mathematical models requires an equation for the conservation of inhibitor. This complicates the consequent mathematical models and their development considerably.

The slow-binding inhibitors bind the enzyme so slowly that the degree of inhibition increases during the measurement of initial velocity. Thus the most serious problem presented by these inhibitors is the experimental one of how to measure both the enzymatic reaction rate and the inhibition rate. The elementary solution to this problem is to incubate the inhibitor with the enzyme until the inhibition is complete and start the reaction with the addition of substrate. Although this approach might be satisfactory with noncompetitive inhibitors, the substrate will change the degree of inhibition by competitive and uncompetitive ones. In addition valuable information about the process of inhibition itself will be lost. Therefore, the processes of inhibition and reaction are measured simultaneously by the measurement and analysis of reaction progress curves. Although these are not initial-velocity measurements, they are included in the present book because of their importance.

Although the examples of slow inhibitors or tight inhibitors are somewhat limited, a larger group of inhibitors are both slow- and tight-binding inhibitors, which achieve additional importance because many of them are physiological molecules and a number of them are proteins. For example the group of proteins called serpins inhibit serine protease enzymes. They include antithrombin III, α_2-antiplasmin, plasminogen activator inhibitors, and α_1-antichymotrypsin.

In the sections that follow the consequences of each set of complications will be discussed separately before those of the combination of the two. For a more detailed presentation of these three types of inhibitors than that below the reader is referred to articles by Cha [1], by Morrison and Walsh [2] and by Williams and Morrison [3].

9.2. Tight-Binding Inhibitors

The spectrum of tightness from the highly reversible inhibitors to irreversible inhibitors was discussed in Chapter 6. It will be useful to discuss irreversible inhibitors further in order to contrast them with tight-binding inhibitors.

9.2.1. IRREVERSIBLE INHIBITORS

Irreversible inhibitors are really peripheral to the purposes of this book and the reader is referred to more detailed discussions elsewhere (*e.g.* [4]). Experimentally the process of inhibition can be studied. In the most common experiment the enzyme and the inhibitor are incubated together in appropriate concentrations so that samples can be removed at various times for the measurement of residual enzyme activity. The conditions are arranged so that the inhibitor is sufficiently dilute during the measurement of enzyme activity that no significant further inhibition occurs. The kinetics of the decay of enzyme activity are generally first order in active enzyme concentration. If the rate of decay is compared with that from the same experiment in the presence of substrate, or more commonly a substrate analog, evidence can be accumulated to decide whether the irreversible inhibitor is competitive or noncompetitive. The presence of substrate, or substrate analog, will diminish the rate of inhibition by a competitive, irreversible inhibitor, whereas it will not affect the rate of inhibition by a noncompetitive, irreversible inhibitor, because the former binds and commonly forms a covalent bond at the active site of the enzyme. Therefore, the competitive, irreversible inhibitors have been extremely valuable in the identification of the active site region of enzymes. Therefore, they are also called active site reagents.

A subpopulation of the latter inhibitors are the suicide inhibitors that bind to the active site of the enzyme and undergo at least part of the catalytic cycle, by which they are activated to form a covalent bond with the enzyme. Suicide inhibitors have been especially valuable for the elucidation and confirmation of the chemical mechanism of numerous enzymes, as well as for pharmacologic agents.

9.2.2. REVERSIBLE, TIGHT-BINDING INHIBITORS

Tight-binding, reversible inhibitors have dissociation constants between irreversible inhibitors and the highly reversible ones. Tight-binding inhibitors are those of which a significant fraction of the total inhibitor binds to the enzyme in order to observe intermediate levels of inhibition. For example if the concentration of enzyme that catalyzes a measurable rate of reaction is 10^{-9} molar, and if the dissociation constant for an inhibitor is also 10^{-9} molar, the total concentration of inhibitor at which 50% inhibition will be measured is 1.5×10^{-9} molar. Under these conditions two thirds of the inhibitor will be free and one third will be bound to the enzyme. Therefore, the concentration of free inhibitor is significantly different from the concentration of total inhibitor. The definition of a tight-binding inhibitor also depends upon the concentration of enzyme necessary to measure the initial velocity. For example if an appropriate reaction rate can be observed in the presence of 10^{-11} molar enzyme, the inhibitor above is no longer a tight-binding inhibitor, because in the latter case only about 0.5% of the total inhibitor will be bound to the enzyme at 50% inhibition.

However, the definition is generally accepted to include those inhibitors with a dissociation constant around 10^{-9}.

Experiments and Results
The same initial-velocity experiments can performed with these tight-binding inhibitors as were done with the readily reversible, analog inhibitors discussed in Chapter 6 except that the concentration of inhibitor is much lower. Ideal data consists of substrate concentration and inhibitor concentration as independent variables with initial velocity as the dependent variable.

Data Analysis and Models
Except for the low dissociation constant and the complicated algebra, presented below, the results of experiments with tight-binding inhibitors can be analyzed is the same ways as those from highly reversible inhibitors [5]. Data analysis from experiments in which the inhibitor concentration and the substrate concentrations are varied over a sufficiently wide range can distinguish competitive, uncompetitive and noncompetitive inhibition, but that has rarely been done.

The complexities are most easily demonstrated for a noncompetitive model. The

initial velocity models are
the same as discussed in
Chapter 6, and the chemical
model for a noncompetitive
i n h i b i t o r   o f
a   unireactant   enzyme
(Figure 9.1) is the same, and
gives   rise   to   the
mathematical   model
(equation 9.1), in which the
two dissociation constants
for inhibitor are assumed to
be equal.

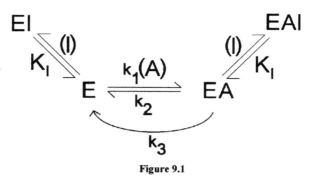

Figure 9.1

However, the term for inhibitor concentration in the mathematical models must
be replaced by a more complicated expression. The conservation equation for enzyme

$$v_i = \cfrac{1}{\left[\cfrac{1}{(A)} * \cfrac{K_A}{V_{max}} + \cfrac{1}{V_{max}} \right] \left[1 + \cfrac{(I)}{K_I} \right]}$$

(9.1)

relates the concentration of total enzyme (E_0) to that of two inhibited forms (E_I) and that
of two catalytically active forms (E_C), (equation 9.2). The conservation equation for
inhibitor relates total inhibitor concentration, the concentration added to the reaction
mixture, to the concentration of free and bound inhibitor (equation 9.2).

$$(E_0) = (EI) + (EAI) + (E) + (EA) = (E_I) + (E_C)$$
$$(I_0) = (I) + (EI) + (EAI) = (I) + (E_I)$$

(9.2)

The two dissociation equations for inhibitor can be added together to express a global
dissociation constant (equation 9.3).

$$K_I = \frac{(E)*(I)}{(EI)}$$

$$K_I = \frac{(EA)*(I)}{(EAI)}$$

$$K_I*(EI) = (E)*(I) \tag{9.3}$$

$$K_I*(EAI) = (EA)*(I)$$

$$K_I = \frac{[(E)+(EA)]*(I)}{(EI)+(EAI)} = \frac{(E_C)*(I)}{(E_I)}$$

The conservation equations are substituted into the equation for the global dissociation constant for the enzyme-inhibitor complex which relates the concentration of free enzyme and free inhibitor (equation 9.4).

$$K_I = \frac{[(E_0)-(E_I)]*[(I_0)-(E_I)]}{(E_I)} \tag{9.4}$$

Solution of the resulting equation for the concentration of bound inhibitor (or inhibited enzyme) requires finding the roots of a quadratic equation (equation 9.5).

$$0 = (E_0)*(I_0) - [(E_0)+(I_0)+K_I]*(E_I)+(E_I)^2$$

$$(E_I) = \frac{(E_0)+(I_0)+K_I \pm \sqrt{[(E_0)+(I_0)+K_I]^2 - 4*(E_0)*(I_0)}}{2} \tag{9.5}$$

The latter equation can be substituted back into the conservation equation for inhibitor (equation 9.2) to get an expression (equation 9.6) for the concentration of free inhibitor for use in the mathematical model for inhibition (equation 9.1).

$$(I) = (I_0) - \frac{(E_0)+(I_0)+K_I \pm \sqrt{[(E_0)+(I_0)+K_I]^2 - 4*(E_0)*(I_0)}}{2} \tag{9.6}$$

The previous set of equations describes a noncompetitive model in which the inhibitor dissociation constants from the enzyme-inhibitor complex and the enzyme-substrate-inhibitor complex are equal, although the corresponding model in which they are different is much more complicated. Furthermore, the set of equations for the competitive and uncompetitive models are also substantially different and more complex. In both of the latter models a change in substrate concentration will change the distribution between free enzyme (E) and the enzyme substrate complex (EA) and therefore, cause a change in the distribution of inhibitor between free inhibitor (I) and bound inhibitor (e.g. (EI) for a competitive model) and, therefore, can additionally change the degree of inhibition. Although the approach above for the development of a mathematical model for either of the latter two chemical models leads to a cubic equation for the concentration of free inhibitor, Morrison [5] developed a generalized approach to a quadratic equation for the initial velocity of enzymes in the presence of tight-binding inhibitors, including competitive and uncompetitive. However, because of the generalized approach the nomenclature and symbolism are difficult to follow. In addition the derivation is quite long and laborious. Therefore, only the equations and the solutions for the competitive (equation 9.7) and the uncompetitive (equation 9.8) are given here with the symbolism defined above.

$$v_i^2 + k_{cat}*(A)*\left[\frac{(I_t)-(E_0}{(A)+K_A} + \frac{1}{\dfrac{K_A}{K_I}}\right]*v_i - \frac{k_{cat}^2*(A)^2*E_0}{\left[K_A+(A)\right]*\dfrac{K_A}{K_I}} = 0$$

$$v_i = \frac{k_{cat}*(A)}{2}*\left[\sqrt{\left(\frac{I_t-E_0}{(A)+K_A} + \frac{1}{\dfrac{K_A}{K_I}}\right)^2 + \frac{4*E_0}{\dfrac{K_A}{K_I}*\left[K_A+(A)\right]}} - \left(\frac{(I_t)-(E_0}{(A)+K_A} + \frac{1}{\dfrac{K_A}{K_I}}\right)\right]$$

(9.7)

$$v_i^2 + k_{cat}*(A)*\left[\frac{(I_t)-(E_0}{(A)+K_A} + \frac{1}{\dfrac{(A)}{K_I}}\right]*v_i - \frac{k_{cat}^2*(A)^2*E_0}{\left[K_A+(A)\right]*\dfrac{(A)}{K_I}} = 0$$

$$v_i = \frac{k_{cat}*(A)}{2}*\left[\sqrt{\left(\frac{I_t-E_0}{(A)+K_A} + \frac{1}{\dfrac{(A)}{K_I}}\right)^2 + \frac{4*E_0}{\dfrac{(A)}{K_I}*\left[K_A+(A)\right]}} - \left(\frac{(I_t)-(E_0}{(A)+K_A} + \frac{1}{\dfrac{(A)}{K_I}}\right)\right]$$

(9.8)

In addition both Morrison [5] and Cha [1] review the utility of other kinds of experiments and mathematical models to describe these chemical models. Furthermore, some curve-fitting programs will determine numerically the roots of a polynomial of higher order than second (e.g. cubic equation).

9.3. Slow-Binding Inhibitors

Slow-binding inhibitors are also defined in the context of the enzyme system under experimentation. An inhibitor whose effect is manifest over a period of time comparable to that over which the initial velocity is measured can be investigated as a slow inhibitor.

9.3.1. EXPERIMENTS AND RESULTS

Since the true inhibited initial velocity is difficult to measure in the presence of the inhibitor, a time course of the reaction is recorded. Although the reaction may be initiated by either the addition of enzyme or the addition of substrate, the former yields results whose analysis is more familiar. The resulting progress curves (Figure 9.2) at different concentrations of inhibitor will start at a rather rapid rate and gradually become slower as more of the enzyme is inhibited and finally approach a constant, inhibited rate as an asymptote. Similar experiments can be done with several concentrations of substrate, but the data collection becomes sufficiently extensive that usually only the inhibitor concentration is varied extensively with the substrate concentration at a value near its K_M. In any case some sort of electronic data recording becomes almost a necessity.

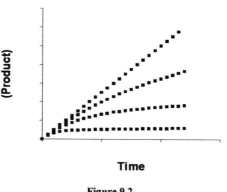

Figure 9.2

9.3.2. DATA ANALYSIS AND MODELS

Data analysis requires a specific model for each of two simultaneous processes: the enzymatic reaction and the inhibition of the enzyme. There is a choice of either of two approaches. Analytic integration of the differential equations with conventional curve fitting or graphic analysis is elegantly described by Morrison and Walsh [2]. Alternatively simultaneous numeric integration and model fitting of the differential equations as described by Taylor et al. [6] is accurate and convenient, once familiarity with the process and the computer program is gained. The former requires less manipulation of computer hardware and software but the number of models that are tested is soon limited by the complexity of analytic integration of each. The former approach will be described first followed by a description of the latter.

There are two dimensions of variability of the chemical models. In the first dimension there are a number of models for the binding of inhibitor to enzyme, but the ones most commonly entertained are the one-step model and a two-step model in which the enzyme and inhibitor form a loose complex that becomes stronger in a second reaction. The second dimension of variability is determined by the enzyme form to which the inhibitor binds, whether the inhibition is noncompetitive, competitive or uncompetitive.

Analytical Integration
One-Step Model. In the simplest model the enzyme reaction will be assumed to conform to a unireactant, Michaelis-Menten model and the inhibition to a noncompetitive, single step chemical model (Figure 9.3). The following discussion requires the assumption that the concentration of substrate changes insignificantly (*c.a.* less than 10%) during the period over which the data is collected. Although this assumption is usually approximated rather well, the model can be supplemented with an equation for the conservation of substrate, if it becomes a problem.

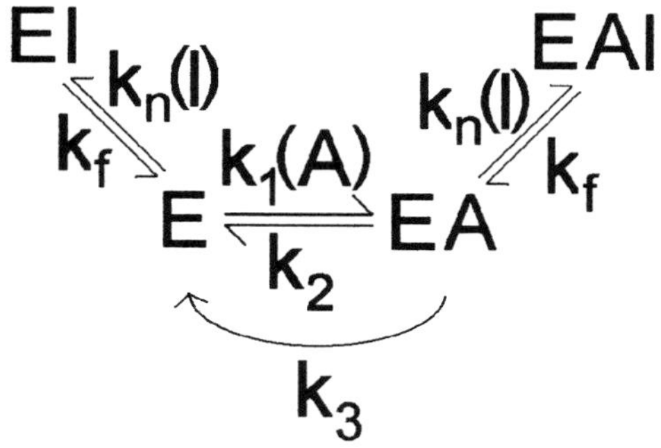

Figure 9.3

The differential equation for the chemical model for the enzyme inhibition (equation 9.9) is for the disappearance of catalytic (uninhibited) enzyme, E_C,

$$-\frac{d(E_C)}{dt} = k_n * (E_C) * (I) - k_f * (E_I)$$

(9.9)

which is the sum of the concentrations of free enzyme, E, and the enzyme-substrate complex, EA. The inhibited enzyme, E_I, is the sum of concentrations of EI and EAI.

The differential equation for the mathematical model for the inhibition (E equation 9.9) is integrated and then combined with the model for the reaction rate, which is integrated again to give a useful expression for product formation as a function of time (equation 9.10).

$$(P) = v_s * t + \frac{[v_0 - v_s] * [1 - \exp[-k * t]]}{k}$$

(9.10)

However, since the process is somewhat long and laborious, the details are relegated to an appendix (Appendix 9.1) to this chapter (section 9.8). The symbol (P) is product concentration, t is time, v_s is the fully inhibited velocity (at infinite time), v_0 is the velocity as soon as the steady-state is established but before significant inhibition; k is the observed rate constant for the process to convert v_0 to v_s.

The fits of time-course data to this equation will result in values for v_s, v_0 and k. For the model above (Figure 9.3) v_0 is independent of inhibitor concentration. The value of v_s is a function of the inhibitor concentration in the same way as a fast, reversible, noncompetitive inhibitor. The value of k is a function of the rate of association, k_n, of the enzyme and the inhibitor, the dissociation rate of the enzyme-inhibitor complex, k_f, (equation 9.11), and the value of C, which is 1.0 for noncompetitive inhibition but is a function of substrate concentration and the apparent K_M value in the models for competitive and uncompetitive inhibition (Appendix 9.1, section 9.8).

$$k = k_n * C * (I) + k_f$$
$$C = 1.0$$

$$(9.11)$$

The fact that the value of k will be a linear function of inhibitor concentration is frequently used as an indicator for a one-step mechanism. The values for the rate constants for inhibition can be calculated from the slope and intercept of the plot of k *vs* inhibitor concentration.

The derivation of the mathematical models for competitive and uncompetitive inhibition only requires the substitution of alternative equations for C, which contain only terms for substrate concentration and the Michaelis constant for the substrate (Appendix 9.1, section 9.8). The more complicated equation for a noncompetitive model with unequal inhibition constants is not given.

Two-Step Model. Mathematical models for two, two-step chemical models for inhibition have been considered [3]. In one of them, discussed below, the inhibitor binds in a rapidly formed complex in one step which then slowly becomes a tighter complex in a second step. In the other chemical model the enzyme first undergoes a slow conformation change to a form that binds the inhibitor in a rapid second step. However, no or few additional inhibition models have been considered. Although evidence has been produced for a number of one-step mechanisms, it is probable that many or all have a preliminary, rapid association step, which is not observed because the dissociation constant is too large.

The most common two-step model considered (Figure 9.4) is one in which the inhibitor first binds and inhibits in a rapid step. Then each initial inhibited complex reacts slowly, associated with a hypothetical conformation change, to a tighter complex. Of course the meaning of rapid and slow depend on the techniques for measurement of

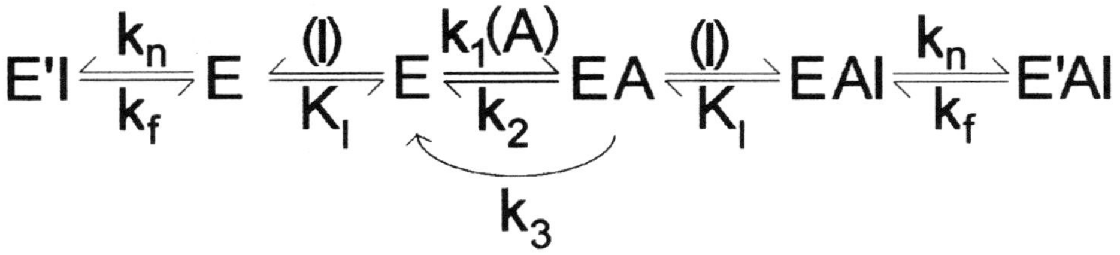

$$E'I \underset{k_f}{\overset{k_n}{\rightleftharpoons}} E \underset{K_I}{\overset{(I)}{\rightleftharpoons}} E \underset{k_2}{\overset{k_1(A)}{\rightleftharpoons}} EA \underset{K_I}{\overset{(I)}{\rightleftharpoons}} EAI \underset{k_f}{\overset{k_n}{\rightleftharpoons}} E'AI$$

$$k_3$$

Figure 9.4

reaction progress. The present enzyme reaction will conform to a unireactant, Michaelis-Menten model and the inhibition to a noncompetitive, two-step chemical model (Figure 9.4). The integrated mathematical model for product formation can be derived for the two-step inhibition in similar manner to that for the one-step chemical model.

Equation 9.12 is the differential equation for the disappearance of the steady-state enzyme, E_C,

$$-\frac{d(E_C)}{dt} = k_n * (E_I) - k_f * (E_I')$$ (9.12)

where E_C is the total steady-state enzyme including the reversibly inhibited species, E_I is the total enzyme inhibited reversibly and E'_I is the total of the enzyme that has been inhibited by the slow process. Integration of equation 9.12, substitution into the differential equation for product formation, and subsequent integration of the latter equation produces a mathematical model (equation 9.13) for product formation as a function of time (Appendix 9.2, section 9.9) whose overall form is identical to the mathematical model for a single-step chemical model (equation 9.10). Therefore, data for each time course of the reaction at different concentrations of inhibitor can be fit to the integrated mathematical model (equation 9.13) to give values of v_0, v_s, and k.

$$(P) = v_s * t + \frac{[v_0 - v_s] * [1 - \exp[-k * t]]}{k}$$ (9.13)

The value of v_0 is the velocity at zero time, but in the present model that velocity is reversibly inhibited noncompetitively by the inhibitor, I. Therefore, v_0 equals the expression for noncompetitive inhibition derived in Chapter 6 and will be an inverse hyperbolic function of inhibitor concentration. It is frequently possible to distinguish data that fits one-step from data that fits two-step models by inspection of a plot of progress curves at different inhibitor concentrations, because the former all begin at the same slope, whereas the latter begin with decreasing slopes with increasing inhibitor concentration. Frequently the value of k (equation 9.14) is plotted *vs* inhibitor concentration to derive values of k_f and k_n. Data that conforms to the present, two-step model will result in a hyperbolic plot whereas data that conforms to a one-step model will result in a linear plot. Values for k_n and k_f can also be calculated from the fit of the model to the data for such a plot. The equation for v_s is somewhat more complicated and the reader is referred to Appendix 9.2 (section 9.9) for its description.

$$k = k_n * C + k_f$$

$$C = \frac{1}{1 + \dfrac{K_I}{(I)}} \tag{9.14}$$

The Mathematical models for competitive and uncompetitive inhibition are the same as the present one, but with alternative expressions for F_A and C (Appendix 9.2).

Numeric Integration
The principal limitation with the above approach to data analysis is that each different chemical model for the inhibition or the enzyme reaction requires reintegration of the differential equations and some of the possible chemical models almost certainly do not have explicit analytical integrals. Therefore, one is limited to those models that can be integrated. Numeric integration of the differential equations is an approach to circumvent these limitations. The differential equations are relatively easy to write, and several programs are available (*e.g.* Scientist® by Micromath) that will simultaneously perform numeric integration to evaluate the dependent variables as a function of time and search for the parameters that give the best fit of these variables to a set of experimental data.

Although different programs will differ from each other with respect to the required format, certain kinds of information must be supplied (Table 9.1).
In addition Appendix 9.3 (section 9.10) contains a listing in the format for Scientist® of a program for the numeric integration of the mathematical model corresponding to the one-step chemical model described above. The programs for competitive and uncompetitive inhibition require the alternative expressions for C. Comparison of the parameter values resulting from numeric integration with those from curve fits with the equation resulting from analytic integration for the same chemical model reveals only an insignificant difference [6].

Table 9.1. Outline for numeric integration of the mathematical model for product formation by a unireactant enzyme in the presence of a slow, single-step, noncompetitive inhibitor.

Section	Example
List independent variables.	time, t
List dependent variables.	product concentration, P
List parameters for which the best values are to be searched.	k_n, k_f, V_{max}, K_A, unless one or more of these is already known.
Give all parameter and concentration values that remain fixed.	E_{0t}, A, I_{0t}
Give conservation equations and any others that are more conveniently done separately.	$EI = E_{0t} - E_t$, $C = 1.0$
Give the differential equations for enzyme and for product in appropriate format.	$(dE_t/dt) = k_n * I * E_t - (k_f/C) * (EI)$ $(dP/dt) = A * V_{max}/(A + K_A)$
Give the initial values for the parameters for which the best values are to be searched.	$k_n = .0018$, *etc.*
Give the initial values of the variables.	$t = 0$, $P = 0$
Give the size of the interval for integration or the ending value of the independent variable.	

9.4. Slow-, Tight-Binding Inhibitors

Most of the inhibitors encountered experimentally that are the subject of this chapter are both slow-binding and tight-binding. Tight-binding inhibitors are almost invariably also slow-binding inhibitors and Cha [1] states an argument why this must be so.

9.4.1. EXPERIMENTS AND RESULTS

Useful experiments to elucidate the mechanism of slow, tight inhibitors are the same as those for slow inhibitors except that the initial concentration of inhibitor is much lower.

9.4.2. DATA ANALYSIS AND MODELS

The data analysis is very much simplified, if the experimental conditions are optimized to approximate as closely as possible each of two assumptions. The initial substrate concentration should be at least ten times the concentration of product formed at the end of the useful time course. The appearance of product should be measured until the rate becomes sufficiently close to the asymptotic value (v_s) so that the latter can be evaluated with acceptable precision. The success of this approximation depends upon the sensitivity of the assay method and the cost of the substrate. Second, the initial inhibitor concentration should be at least ten times that of the enzyme. The success of this approximation depends upon the turnover number, k_{cat}, of the enzyme, the sensitivity with which product can be measured and the overall dissociation constant for the enzyme-inhibitor complex. Frequently the time course of the reaction must be measured with a highly inhibited enzyme. If especially the second of these two approximations can be approximated, the data analysis is the same as that for slow inhibitors in section 9.3. Because of the high inhibitor levels the distinction of competitive from uncompetitive and noncompetitive is frequently not made, particularly if all of the experiments are done at a single concentration of substrate.

The relatively high levels of inhibitor are sometimes more in the range of the dissociation constant of the first highly reversible binding in the two-step model. Therefore, the approximation is somewhat better with the latter chemical model.

However, if the approximations cannot be realized, the analysis becomes more complicated. In these situations it is necessary to somehow include an equation for the conservation of substrate, an equation for the conservation of inhibitor or both. The complications due to an equation for the conservation of substrate are not insurmountable. Unfortunately with an equation for the conservation of inhibitor analytical integration of the differential equations is possible only for the simplest models, *e.g.* a one-step model, and even that requires considerable labor and results in some uncertainty, because of the necessity to solve polynomial equations.

Therefore it is almost imperative to integrate these models numerically. Numeric integration of the single-step models is very similar to that in Appendix 9.3, with the appropriate expression for C and the addition of an equation for the conservation of inhibitor. However, the analogous program for the corresponding two-step model, in which the inhibitor rapidly forms an enzyme complex, can be written to include numeric integration of the differential equation for the initial velocity in the presence of a tight-binding inhibitor (equation 9.7 and equation 9.8) in place of the inhibited Michaelis-Menten equation ordinarily used. In addition the concentration of the reversibly inhibited enzyme complex can also be calculated from the same quadratic equation. However, examples of this approach are very few.

9.5. Examples

The study of slow-, tight-binding inhibitors is illustrated with two examples, one a physiologic molecule and the other a synthetic one.

The anticoagulant protein anophelin is isolated from the salivary gland of the mosquito. It is a tight-, slow-binding inhibitor of α-thrombin. Data from reaction progress curves with a chromogenic substrate [7] were fit to equation 9.10 and the values of k were linear with total inhibitor concentration. The linearity supported the hypothesis of an inhibition mechanism in which only a single step could be demonstrated, and the rate constants were calculated from the slope and intercepts of the line. The inhibition conformed to a competitive chemical model, although additional models for the inhibition were not tested.

The inhibition of hepatitis C virus protease by a hexepeptide-α-ketoacid conformed to a hypothetical two-step chemical mechanism [8] in which a rapidly formed complex slowly isomerizes to a tighter complex. The hypothesis is supported by the hyperbolic nature of the plot of the observed rate constant for the slow inhibition, k, vs inhibitor concentration and the inverse hyperbolic nature of plot of the initial velocity, v_0, vs the inhibitor concentration. The lowest inhibitor concentration in the progress curves (5.0 nM) was well above the dissociation constant for the tight complex (0.05 nM) in order to approximate the assumption that the free inhibitor concentration was the same as the total inhibitor concentration.

9.6. Summary

In summary there are tight-binding inhibitors, slow-binding inhibitors and tight-, slow-binding inhibitors. The latter are frequently physiological inhibitors. The analysis of data with tight-binding inhibitors is complicated by the requirement for equations for the conservation of inhibitor.

The discrimination of models for slow-binding inhibitors requires the integration of equations for enzyme inhibition and for product formation. The analysis of data with tight-, slow-binding inhibitors requires both of the above, unless successful approximations can be made to two assumptions: that the concentration of neither inhibitor nor substrate changes during the course of the data collection. Numeric integration of the models that include slow-binding inhibitors is not only easier but also permits the testing of more chemical models and rigorous elimination of the assumptions for tight-, slow-binding inhibitors in some cases.

9.7. References

1. Cha, S. "Tight-Binding Inhibitors-1," *Biochem. Pharmacol.* 24, 2177-85 (1975).

2.Morrison, J.F. and Walsh, C.T. "The Beharior and Significance of Slow-Binding Enzyme Inhibitors," *Adv. Enzymol. Relat. Areas Mol. Biol.* 61, 201-301(1988).

3. Williams, J. and Morrison, J.F. "The Kinetics of Reversible Tight-Binding Inhibition," *Methods Enzymol.* 63, 437-67 (1979).

4. Plapp, B.V. "Application of Affinity Labeling for Studying Structure and Function of Enzymes," *Methods Enzymol.* 87, 469-99 (1982).

5. Morrison, J.F. "Kinetics of the Reversible Inhibition of Enzyme-Catalyzed Reactions by Tight-Binding Inhibitors," *Biochim. Biophys. Acta*, 185, 269-86.

6. Taylor, K.B., Windsor, L.J., Caterina, N.C., Bodden, M,K. and Engler, J.A. "The Mechanism of Inhibition of Collagenase by TIMP-1," *J. Biol. Chem.* 271, 23938-45, (1996).

7. Francischetti, I.V.M., Valenzuela, J.G. and Ribeiro, J.M.C. "Anophelin: Kinetics and Mechanism of Thrombin Inhbition," *Biochemistry*, 38, 16678-85 (1999).

8. Narjes, F. Brunetti, M., Colarusso, S., Gerlach, B., Koch, U., Biasiol, G., Fattori, D., De Francesco, R., Matassa, V.G. and Steinkühler, C. "α-Ketoacids Are Potent Slow Binding Inhbitors of the Hepatitis C Virus NS3 Protease," *Biochemistry*, 39, 1849-61 (2000).

9.8. Appendix 9.1: Derivation of Mathematical Model of Slow, Noncompetitive, One-Step Inhibition of a Unireactant Enzyme

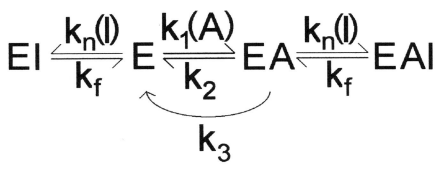

Figure 9.5

The other symbols below are:

> E_0=total enzyme at zero time
> E_I=total inhibited enzyme at any time
> E_C=total active (catalytic) enzyme at any time

1. The equation for the conservation of enzyme shows that the enzyme can be divided at any time into catalytic enzyme, E_C, and inhibited enzyme, E_I.

$$(E_C)=(E)+(EA)$$
$$(E_I)=(EI)+(EAI)$$
$$(E_0)=(E)+(EA)+(EI)+(EAI)$$
$$(E_0)=(E_C)+(E_I)$$

(9.15)

2. The differential equation for the rate of progression of inhibition (the rate of formation of inhibited enzyme) equals the negative of that for the reaction of catalytic enzyme with inhibitor (equation 9.16).

$$\frac{d(E_I)}{dt}=-\frac{d(E_C)}{dt}=k_n*(E_C)*(I)-k_f*(E_I)$$

(9.16)

3. With substitution of the equation for the conservation of enzyme this becomes equation 9.17.

$$-\frac{d(E_C)}{dt}=k_n*(E_C)*(I)-k_f*(E_0)-k_f*(E_C)$$

(9.17)

4. In noncompetitive inhibition, the inhibitor reacts with both forms of the catalytic enzyme. This is not true of competitive and uncompetitive inhibition presented later. Therefore, a factor, C, will be defined as the fraction of the catalytic enzyme with which the inhibitor reacts. For noncompetitive inhibition:

$$C=\frac{(E)+(EA)}{(E_C)}$$

$$C=1.0$$

(9.18)

5. Substitute the relationship in equation 9.18 into the differential equation (equation 9.17) and collect those terms with similar multipliers.

$$-\frac{d(E_C)}{dt}=k_n*(E_C)*C*(I)-k_f*(E_0)+k_f*(E_C)$$

$$-\frac{d(E_C)}{dt}=\left[k_n*C*(I)+k_f\right]*(E_C)-k_f*(E_0)$$

(9.19)

$$-\frac{d(E_C)}{dt}=k*(E_C)+a$$

where

$$k=k_n*C*(I)+k_f$$

$$a=-k_f*(E_0)$$

(9.20)

6. Integration of the following expression between E_0 and E_C; and between 0 and t will give the equation for enzyme inhibition.

$$\int \frac{1}{k*(E_C)+a}*d(E_C)=\int -dt \tag{9.21}$$

7. The result of this integration is:

$$(E_C)=\frac{[a+k*E_0]*\exp[-k*t]-a}{k}$$

$$(E_C)=\left[\frac{a}{k}+E_0\right]*\exp[-k*t]-\frac{a}{k} \tag{9.22}$$

8. The velocity of the enzyme reaction is:

$$v=\frac{(E_C)*k_{cat}*(A)}{(A)+K_M}=(E_C)*F_A \tag{9.23}$$

9. Substitution of the equation for the concentration of catalytic enzyme (equation 9.22) into the latter equation (equation 9.23) results in:

$$v_i=\left[\frac{F_A*a}{k}+F_A*E_0\right]*\exp[-k*t]-\frac{F_A*a}{k} \tag{9.24}$$

10. The velocity at t=0 (v_0) and at full inhibition (v_s) (infinite time) respectively are:

$$v_0=F_A*(E_0)$$

$$v_s=-F_A*\frac{a}{k} \tag{9.25}$$

11. Substitution back into equation 9.24 and some rearrangement results in equation 9.26.

$$v=\frac{d(P)}{dt}=v_s-[v_s-v_0]*\exp[-k*t] \tag{9.26}$$

12. Integration of this differential equation between the limits of zero to (P) and zero to t results in the equation for the concentration of product as a function of time.

$$(P) = v_s * t + \frac{[v_0 - v_s] * [1 - \exp[-k * t]]}{k}$$

(9.27)

13. For uncompetitive inhibition:

$$C = \frac{(EA)}{(E_C)}$$

$$C = \frac{(A)}{(A) + K_A}$$

(9.28)

14. For competitive inhibition:

$$C = \frac{(E)}{(E_C)}$$

$$C = \frac{K_A}{(A) + K_A}$$

(9.29)

9.9. Appendix 9.2: Derivation of Mathematical Model of Slow Inhibition, Unireactant Enzyme, Noncompetitive, two-Step Inhibition

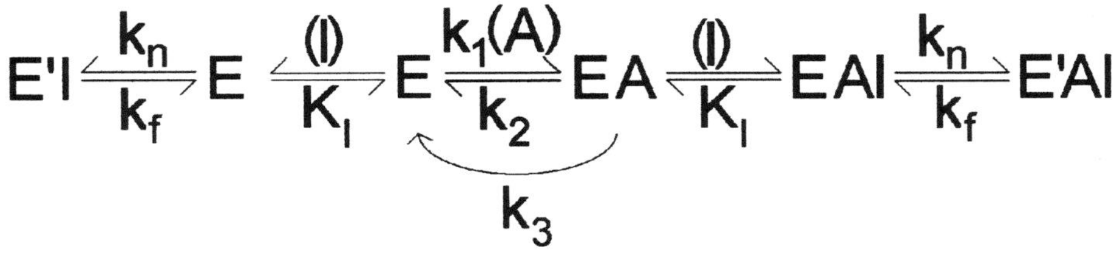

$$E'I \underset{k_f}{\overset{k_n}{\rightleftharpoons}} E \underset{K_I}{\overset{(I)}{\rightleftharpoons}} E \underset{k_2}{\overset{k_1(A)}{\rightleftharpoons}} EA \underset{K_I}{\overset{(I)}{\rightleftharpoons}} EAI \underset{k_f}{\overset{k_n}{\rightleftharpoons}} E'AI$$

$$k_3$$

Figure 9.6

The symbols below are:

E_0=total enzyme at zero time

E_C=total steady-state enzyme at any time

E_I=steady-state inhibited enzyme at any time

E'_I=slowly-inhibited enzyme at any time

1. The equation for the conservation of enzyme is:

$$(E_C) = (E) + (EA) + (EI) + (EAI)$$

$$(E_I) = (EI) + EAI) \tag{9.30}$$

$$(E_0) = (E) + (EA) + (EI) + (EAI) + (E'I) + (E'AI) = (E_C) + (E_I)$$

2. The differential equation for the reaction of the steady-state, enzyme-inhibitor complex to form the slow complex is:

$$\frac{d(E_I')}{dt} = -\frac{d(E_C)}{dt} = k_n*(E_I) - k_f*(E_I') \tag{9.31}$$

3. With substitution of the equation for the conservation of enzyme this becomes:

$$-\frac{d(E_C)}{dt} = k_n*(E_I) - k_f*\left[(E_0) - (E_C)\right] \tag{9.32}$$

4. In noncompetitive inhibition the inhibitor binds to both forms of the catalytic enzyme. This is not true for competitive or uncompetitive inhibition presented later. Therefore, a factor, C, will be defined as the fraction of the steady-state enzyme (E_C) that is rapidly bound to the inhibitor. For noncompetitive inhibition the dissociation constants of the inhibitor from the enzyme-inhibitor complex and from the enzyme-substrate-inhibitor complex can be expressed in a global sense by addition of the two equilibrium expressions.

$$K_I = \frac{(I)*[(E)+(EA)]}{(E_I)}$$

$$\frac{(E_I)}{(E_C)} = \frac{(E_I)}{(E)+(EA)+(EI)+(EAI)}$$

$$\frac{(E_I)}{(E_C)} = \frac{(E_I)}{(E_I)+\dfrac{K_I}{(I)}*(E_I)} = \frac{1}{1+\dfrac{K_I}{(I)}} \tag{9.33}$$

$$C = \frac{1}{1+\dfrac{K_I}{(I)}}$$

$$(E_I)=(E_C)*C$$

5. Substitute the relationship in equation 9.33 into the differential equation (equation 9.32) in order to convert all of the terms for free enzyme-inhibitor complex (EI) to those for total steady-state enzyme (E_C).

$$-\frac{d(E_C)}{dt}=k_n*C*(E_C)-k_f*(E_0)+k_f*(E_C)$$

$$-\frac{d(E_C)}{dt}=\left[k_n*C+k_f\right]*(E_C)-k_f*(E_o) \tag{9.34}$$

$$-\frac{d(E_C)}{dt}=k*(E_C)+a$$

Where:

$$k = k_n * C + k_f$$
$$a = -k_f * (E_0)$$

(9.35)

6. Integration of the following expression between E_0 and E_C; and between 0 and t will give the equation for enzyme inhibition.

$$\int \frac{1}{k*(E_C)+a} * d(E_C) = \int -dt$$

(9.36)

7. The result of this integration is:

$$(E_C) = \frac{[a + k*E_0] * \exp[-k*t] - a}{k}$$

$$(E_C) = \left[\frac{a}{k} + E_0\right] * \exp[-k*t] - \frac{a}{k}$$

(9.37)

8. The velocity of the enzyme reaction in the presence of a steady-state, noncompetitive inhibitor is:

$$v = \frac{(E_C)*k_{cat}*(A)}{[(A)+K_A]*\left[1+\frac{(I)}{K_I}\right]} = (E_C)*F_A$$

(9.38)

where:

$$F_A = \frac{k_{cat}*(A)}{[(A)+K_A]*\left[1+\frac{(I)}{K_I}\right]}$$

(9.39)

The corresponding expressions for competitive and uncompetitive inhibitors respectively are:

$$F_A = \frac{k_{cat} * (A)}{(A) + K_A * \left[1 + \dfrac{(I)}{K_I}\right]} = (E_C) * F_A$$

$$F_A = \frac{k_{cat} * (A)}{(A) * \left[1 + \dfrac{(I)}{K_I}\right] + K_A} = (E_C) * F_A$$

(9.40)

9. substitution of the equation for the concentration of steady-state enzyme produces:

$$v = \left[\frac{F_A * a}{k} + F_A * E_0\right] * \exp[-k * t] - \frac{F_A * a}{k}$$

(9.41)

10. The velocity at t=0 (v_0) and that (v_s) at full inhibition (infinite time) respectively are:.

$$v_0 = F_A * (E_0)$$

$$v_s = -F_A * \frac{a}{k}$$

(9.42)

11. Therefore:

$$v = v_s - \left[v_s - v_0\right] * \exp[-k * t]$$

$$\frac{d(P)}{dt} = v_s - \left[v_s - v_0\right] * \exp[-k * t]$$

(9.43)

12. Integration of this differential equation between the limits of zero to (P) and zero to t results in an expression for product concentration as a function of time.

$$(P) = v_s * t + \frac{\left[v_0 - v_s\right] * \left[1 - \exp[-k * t]\right]}{k}$$

(9.44)

13. For competitive inhibition:

$$C = \frac{\dfrac{K_A}{(A)} * \left[1 + \dfrac{(I)}{K_I}\right]}{1 + \dfrac{K_A}{(A)} * \left[1 + \dfrac{(I)}{K_I}\right]}$$ (9.45)

14. For uncompetitive inhibition:

$$C = \frac{\dfrac{(I)}{K_I}}{1 + \dfrac{(I)}{K_I} + \dfrac{K_A}{(A)}}$$ (9.46)

9.10. Appendix 9.3. Numeric Integration of Mathematical Model of Slow, Noncompetitive, Single-Step Inhibition of a Unireactant Enzyme: format for Scientist® (Micromath)

	Section Identification
//MicroMath Scientist Model File	
IndVars: t	Independent variables
DepVars: P	Dependent variables
Params: KA, VA, kn, kf	Parameters
I=0.5	Fixed values
Et0=5	
A=0.2	
C=1.0	Other equations
Ec'=(kn*C*I+kf)*Ec-(kf*E0)	
P'=A*Vm/(A+KA)	Differential equations
End of Equations	
//Parameter values	
KA=0.1	
Vm=22	Initial values of parameters
kn=0.018	
kf=0.018	
//Initial conditions	
t=0	
P=0	Initial conditions
//Interval Size	
IntervalSize=8	
//Number of Output points	Interval size
NumOfPoints=100	
*****ENDMODEL	Number of output points

CHAPTER 10

THE THERMODYNAMICS OF INITIAL VELOCITY

10.1. Introduction

The remainder of this book consists of discussions of environmental effects on the initial velocity, *e.g.* the effects of pH, isotope effects, and the effects of temperature. The interest is in both a conceptual as well as a mathematical model for the relationship of the intermediates and transition states in a generalized enzyme-catalyzed reaction to the initial velocity and its operational parameters. Both the conceptual understanding and the interpretation of data from these effects is aided by a discussion of the relation on the one hand of initial velocity and its operational parameters (k_{cat} and k_{cat}/K_M) with the rates of the various processes that together constitute the reaction pathway and on the other hand by a discussion of the effects of each of the environmental conditions above on the rates of the various processes.

  The present chapter contains an exploration of the relationship of the initial velocity of an enzymatic reaction to the energy of the various intermediates and transition states in the pathway. It also contains an exploration of the relationship of a change in the energy of the intermediates and transition states to a change in the initial velocity and its operational parameters, k_{cat}/K_M and k_{cat}. The following chapters contain further discussions of the latter relationships and how they may be investigated in specific experiments.

  Since reaction rates are related to the magnitude of energy transitions, the present discussion will begin with the energy transitions that determine the initial velocity of enzymatic reactions. Then the quantitative effect of changes in these energy transitions on the initial velocity as well as in its operational parameters will be explored.

  Much of the following discussion is in the context of a unireactant enzyme reaction with one enzyme-substrate intermediate and one enzyme-product intermediate. This is an adequate and complete chemical model for some enzyme reactions. It may also be a useful model for the second substrate to bind in a bireactant reaction and the third substrate to bind in a terreactant reaction. In addition the model is useful, with some limitations, for either substrate in a random reaction in the presence of a saturating concentration of the other substrate. Therefore, the illustration of the theory with this simple model is more useful and certainly more easily comprehensible than the rigorous use of a generalized model. Some of the alterations in the mathematical model associated with other chemical models will be discussed.

10.2. Thermodynamics

10.2.1. BASIC CONCEPTS

A simple one-step, nonenzymatic reaction
conceptually consists of reactants and
products each with a given level of energy.
According to reaction-rate theory each
molecule of reactant, A, must change to a
higher-energy form called the transition
state, which with a certain probability will
lose energy to become a molecule of
product, P. These phenomena are
represented graphically in the reaction
diagram (Figure 10.1) in which the vertical

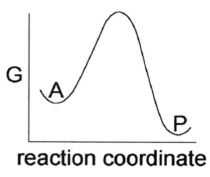

reaction coordinate

Figure 10.1

axis is the energy level and the horizontal axis is called the reaction coordinate. For the
simple association/dissociation reactions of single atoms the reaction coordinate is a bond
distance. However, even in the latter case it is frequently not a single straight line and is
certainly much more complicated in a reaction of multiatomic molecules. However, the
diagram is a convenient concept and it will be useful here.

An enzyme-catalyzed reaction, like other multi-step chemical reactions is
conceived as a sequential series of chemical intermediates with transition states between
them (*e.g.* Figure 10.2). For the sake of discussion unless described to the contrary it will
be generally assumed that there is only a single pathway for each reaction. During the
steady state the intermediates are related to each other by equilibrium constants. For

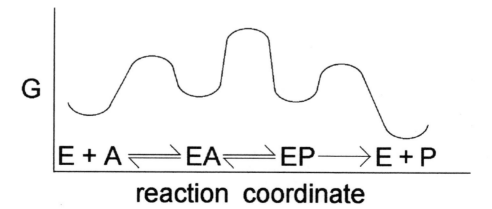

reaction coordinate

Figure 10.2

example the mathematical model in equation 10.1 was derived for the chemical model in Figure 10.2 (also in Figure 10.3). If the downstream equilibrium constant is extremely large the step is regarded as irreversible. Each intermediate can be converted to the next one downstream in a reaction with a rate constant, and, if that step is not irreversible, the downstream intermediate is converted to the upstream intermediate in a reaction with the reverse rate constant.

$$\frac{v_i}{E_t} = \frac{1}{\dfrac{1}{(A)} * \left[\dfrac{1}{k_1} + \dfrac{1}{K_1 * k_3} + \dfrac{1}{K_1 * K_3 * k_5}\right] + \dfrac{1}{k_3} + \dfrac{1}{K_3 * k_5} + \dfrac{1}{k_5}}$$

(10.1)

$$K_1 = \frac{k_1}{k_2} \qquad\qquad K_3 = \frac{k_3}{k_4}$$

Although the reaction profile above is a very useful concept for a generalized enzyme reaction, the development of such a profile for a specific enzyme requires the resolution of several issues.

Intermediates

In spite of the ability to draw a reaction profile it must be recognized that one or more of the intermediates may be more abstract than real. Although the enzyme forms that have been discussed in the previous pages of this book certainly exist, whether or not each of them is a true intermediate constitutes a group of difficult and important questions. In order to be an intermediate enzyme form it must exist in a significant concentration during the steady state, have more than a fleeting existence. Therefore, in the reaction profile it must be represented by a valley in the curve and be bounded on either side by a higher-energy transition state. In addition the nature of each of the putative intermediates is complicated by the fact that several functions may be going on at the same time. For example there is good evidence, *e.g.* from transition state analogs discussed below, that the binding of substrate frequently, if not always, involves some progress in catalysis. It follows then that the catalysis probably involves some release of product. A third source of ambiguity with regard to the intermediates in any specific enzyme reaction originates from the uncertainty in the number of intermediates and transition states that are not detected in steady-state experiments. In general intermediates that do not involve substrate binding or product release will not be detected in steady-state experiments. There may be any number of conformation changes, partial catalytic changes, proton rearrangements, *etc.* that will be transparent in steady-state experiments. Nevertheless certain specific enzyme forms certainly exist during the steady state as intermediates at least long enough to bind to analog inhibitors and product inhibitors. In addition information can be gained about the nature of some of these intermediates from the results of experiments on environmental effects on the initial velocity.

The relationship between the energy and equilibrium is given by equation 10.2. Where R is the gas constant and T is the absolute temperature. Of course, if the downstream step is irreversible, both the equilibrium constant and the energy change are quite large and the latter is negative.

$$K_{eq} = \exp\left[\frac{-\Delta G^\circ}{R * T}\right]$$

(10.2)

Transition States

The transition states in specific enzyme reactions have been described in several different experimental contexts. Certain competitive inhibitors, called transition-state analogs, have structural features that resemble an hypothetical transition state for an enzyme reaction and bind to the enzyme more tightly, *i.e.* with a lower dissociation constant, than the substrate. Much has been learned about the transition state in enzyme reactions by experiments with transition-state analogs. For example the structure of the analog has been regarded as evidence for the structure of the transition state. In addition the binding energy of the analog has been used to estimate the amount of energy available to distort the structure of the substrate upon binding.

The energy profile has been described in detail for the enzyme triose isomerase [1] from the results of a series of initial-velocity and isotope-exchange experiments. However, the approach is applicable only to enzymes the catalyze a reaction in which one or more of the intermediates exchange isotope with the solvent.

The relation between energy and rate of a single-step reaction is given by equation 10.3, Where $\Delta G^\ddagger$ is the energy change from the reactant (or intermediate) to the transition state, where R is the gas constant, T is the absolute temperature, h is Planck's constant, and

$$k = \frac{k_b * T}{h} * \exp\left[\frac{-\Delta G^\ddagger}{R * T}\right]$$

(10.3)

k_b is the Boltzman constant.

10.2.2. INITIAL VELOCITY

As an example of an enzymatic reaction the chemical model for a unireactant enzyme with two intermediates (Figure 10.3) will result in the same mathematical model presented in Chapter 4 for the initial velocity (equation 10.4). For the purpose of illustration take one term from the latter equation and substitute equation 10.2 and 10.3 for the energy relationships for the rate and equilibrium constants (equation 10.5).

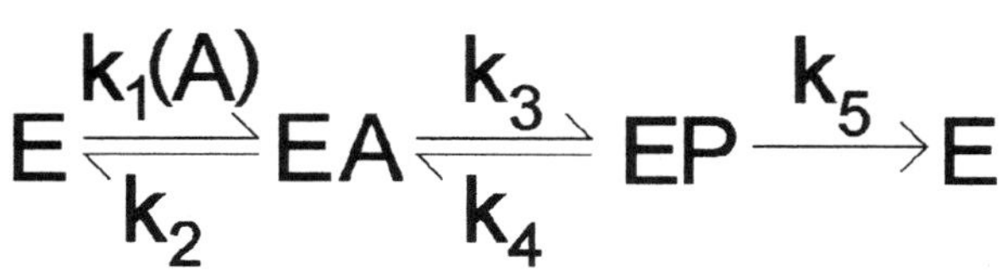

$$E \underset{k_2}{\overset{k_1(A)}{\rightleftharpoons}} EA \underset{k_4}{\overset{k_3}{\rightleftharpoons}} EP \overset{k_5}{\longrightarrow} E$$

Figure 10.3

Repetition of this

$$\frac{v_i}{E_t} = \cfrac{1}{\cfrac{1}{(A)} * \left[\cfrac{1}{k_1} + \cfrac{1}{K_1 * k_3} + \cfrac{1}{K_1 * K_3 * k_5} \right] + \cfrac{1}{k_3} + \cfrac{1}{K_3 * k_5} + \cfrac{1}{k_5}} \qquad (10.4)$$

$$\frac{1}{K_1 * K_3 * k_5} = \cfrac{1}{\cfrac{k_b * T}{h} * \exp\left[\cfrac{-\Delta G_1^{\circ}}{R*T}\right] * \exp\left[\cfrac{-\Delta G_3^{\circ}}{R*T}\right] * \exp\left[\cfrac{-\Delta G_5^{\ddagger}}{R*T}\right]}$$

$$= \frac{h}{k_b * T} * \exp\left[\frac{\Delta G_1^{\circ} + \Delta G_3^{\circ} + \Delta G_5^{\ddagger}}{R*T}\right] \qquad (10.5)$$

substitution for all of the terms in equation 10.4 results in an equation for the initial velocity of the reaction in terms of the energy transitions in the reaction (equation 10.6).

$$\frac{v_i}{E_t} = \frac{k_b * T}{h} * \frac{1}{\sum G}$$

$$\sum G = \frac{1}{(A)} * \left[\exp\left[\frac{\Delta G_1^{\ddagger}}{R*T}\right] + \exp\left[\frac{\Delta G_1^{\circ} + \Delta G_3^{\ddagger}}{R*T}\right] + \exp\left[\frac{\Delta G_1^{\circ} + \Delta G_3^{\circ} + \Delta G_5^{\ddagger}}{R*T}\right] \right] + $$

$$\exp\left[\frac{\Delta G_3^{\ddagger}}{R*T}\right] + \exp\left[\frac{\Delta G_3^{\circ} + \Delta G_5^{\ddagger}}{R*T}\right] + \exp\left[\frac{\Delta G_5^{\ddagger}}{R*T}\right] \qquad (10.6)$$

It can be seen that the energy component of each term is the energy difference between some intermediate to some downstream transition state. For example the fifth term in equation 10.6 is the energy difference between the enzyme intermediate form EA and the transition state for the product release (rate constant k_5). The initial velocity is determined by the energy transition between every intermediate and every downstream transition state with which it is connected by reversible equilibria. The concept that only the energy difference from each intermediate to the subsequent downstream transition state determines the initial velocity seems too simple.

It will be remembered that the k_{cat} (or V_{max}) is determined by all those terms that do not contain a term for the concentration of substrate (equation 5.2). Therefore, it will be a function of the energy difference between every intermediate and every downstream transition state with which it is connected by reversible equilibria as long as that energy difference does not include the binding of substrate. The precise energy differences that determine the k_{cat} depend on the steady-state mechanism. In the present model it is a function of the energy differences indicated in equation 10.7. Because of the approximations in initial-velocity experiments, the k_{cat} will always be a function of the transition-state, energy difference in at least one product release step.

$$k_{cat} = \frac{k_b * T}{h} * \frac{1}{\exp\left[\dfrac{\Delta G_3^{\ddagger}}{R*T}\right] + \exp\left[\dfrac{\Delta G_3^{\circ} + \Delta G_5^{\ddagger}}{R*T}\right] + \exp\left[\dfrac{\Delta G_5^{\ddagger}}{R*T}\right]} \tag{10.7}$$

The k_{cat}/K_M (or V_{max}/K_M) (equation 10.8) is determined by those terms that involve the binding of substrate (equation 5.3). Therefore, it will be a function of any energy differences between the intermediate (enzyme form) to which the substrate binds and any of the downstream transition states that are connected by reversible equilibria as well as those between the same transition states and any intermediates upstream to that to which the substrate binds and connected to the latter by reversible steps.

$$\frac{k_{cat}}{K_A} = \frac{k_b * T}{h} * \frac{1}{\exp\left[\dfrac{\Delta G_1^{\ddagger}}{R*T}\right] + \exp\left[\dfrac{\Delta G_1^{\circ} + \Delta G_3^{\ddagger}}{R*T}\right] + \exp\left[\dfrac{\Delta G_1^{\circ} + \Delta G_3^{\circ} + \Delta G_5^{\ddagger}}{R*T}\right]} \tag{10.8}$$

Thermodynamics and Steady-State Models
The energy transitions that determine the value of the steady-state parameters are different for different steady-state models.

Bireactant, Sequential, Ordered Model. A bireactant, ordered chemical model (Figure 10.4) has the mathematical model for the initial velocity in equation 10.9. The horizontal

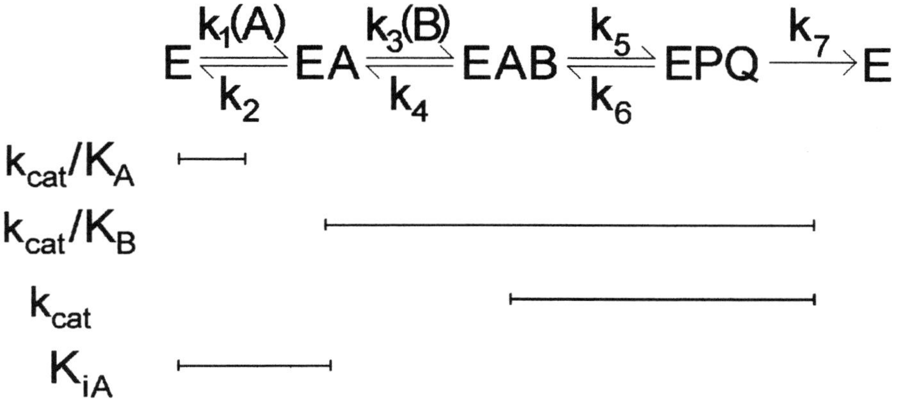

$$E \underset{k_2}{\overset{k_1(A)}{\rightleftharpoons}} EA \underset{k_4}{\overset{k_3(B)}{\rightleftharpoons}} EAB \underset{k_6}{\overset{k_5}{\rightleftharpoons}} EPQ \overset{k_7}{\longrightarrow} E$$

k_{cat}/K_A ⊢——┤

k_{cat}/K_B ⊢————————————————┤

k_{cat} ⊢——————————┤

K_{iA} ⊢————————┤

Figure 10.4

$$\frac{v_i}{E_t} = \frac{1}{D}$$

$$D = \frac{1}{(A)} * \frac{1}{k_1} + \frac{1}{(A)*(B)} * \left[\frac{1}{K_1*k_3} + \frac{1}{K_1*K_3*k_5} + \frac{1}{K_1*K_3*K_5*k_7} \right] + \tag{10.9}$$

$$\frac{1}{(B)} * \left[\frac{1}{k_3} + \frac{1}{K_3*k_5} + \frac{1}{K_3*K_5*k_7} \right] + \frac{1}{k_5} + \frac{1}{K_5*k_7} + \frac{1}{k_7}$$

lines in the figure show the portion of the chemical model that determines each of the operational parameters. According to the definitions in Chapter 5 (equation 5.9) the horizontal line shows that the k_{cat}, or V_{max}/E_t, will be determined by the energy differences from the intermediate, EAB, and the transition state of the catalytic step ($\Delta G_5^{\ddagger}$), that from the same intermediate to the transition state for the product release ($\Delta G_5 + \Delta G_7^{\ddagger}$), and that from the enzyme-product intermediate, EPQ, to the transition state for product release ($\Delta G_7^{\ddagger}$ in equation 10.10).

$$\frac{V_{max}}{E_t} = \frac{1}{\dfrac{1}{k_5} + \dfrac{1}{K_5 * k_7} + \dfrac{1}{k_7}}$$

$$= \frac{k_b * T}{h} * \frac{1}{\exp\left[\dfrac{\Delta G_5^{\ddagger}}{R*T}\right] + \exp\left[\dfrac{\Delta G_5^{\circ} + \Delta G_7^{\ddagger}}{R*T}\right] + \exp\left[\dfrac{\Delta G_7^{\ddagger}}{R*T}\right]}$$ (10.10)

However, it must be realized that any additional, possibly transparent and reversible or irreversible steps that do not involve substrate binding may also be involved, *e.g.* release of additional products.

The value of the k_{cat}/K_A, or V_{max}/K_A*E_t, will be a function only of the energy difference between the free enzyme and the transition state for the binding of the first substrate (equation 5.9 and equation 10.11). Of course the energy difference in any reversible transitions connected with the previously mentioned intermediates that do not involve the binding of the other substrate will also influence this value.

$$\frac{k_{cat}}{K_A} = \frac{1}{\dfrac{1}{k_1}}$$

$$= \frac{k_b * T}{h} * \frac{1}{\exp\left[\dfrac{\Delta G_1^{\ddagger}}{R*T}\right]}$$ (10.11)

The value of the k_{cat}/K_B, or V_{max}/K_B*E_t, will be a function of magnitudes all of the energy transitions starting with the intermediate enzyme form to which the second substrate binds and ending with the transition state for the release of the first product (equation 10.12). Of course the energy difference in any reversible transitions connected with the previously mentioned intermediate that does not involve the binding of the other substrate will also influence this value.

$$k_{cat} = \cfrac{1}{\cfrac{1}{k_3} + \cfrac{1}{K_3 * k_5} + \cfrac{1}{K_3 * K_5 * k_7}}$$

$$= \frac{k_b * T}{h} * \cfrac{1}{\exp\left[\dfrac{\Delta G_3^{\ddagger}}{R * T}\right] + \exp\left[\dfrac{\Delta G_3^{\circ} + \Delta G_5^{\ddagger}}{R * T}\right] + \exp\left[\dfrac{\Delta G_3^{\circ} + \Delta G_5^{\circ} + \Delta G_7^{\ddagger}}{R * T}\right]}$$

(10.12)

The value of the K_{iA} will be a function of magnitude of the energy difference between the free enzyme and the intermediate complex of the enzyme with the first substrate (equation 10.13).

$$K_{iA} = \frac{1}{K_1}$$

$$= \frac{h}{k_b * T} * \exp\left[\frac{\Delta G_1^{\circ}}{R * T}\right]$$

(10.13)

Bireactant, Sequential, Random, Rapid-Equilibrium Model. The initial velocity of a bireactant, sequential, random rapid-equilibrium, model (Figure 10.5) is given in equation 10.14.

The k_{cat} is a function of only k_5, which includes the catalytic step and the release of products (equation 10.15). If the products are also released in rapid-equilibrium steps the k_{cat} will be determined only by the catalytic step, not because of the randomness but because the rapid-equilibrium release of product has a transition state of very low energy. In a random, steady-state chemical model the meaning will be considerably more complicated.

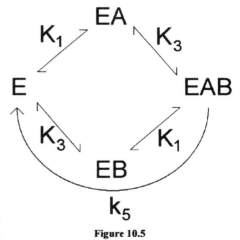

Figure 10.5

$$\frac{v_i}{E_t} = \cfrac{1}{\cfrac{1}{K_1 * k_5} * \cfrac{1}{(A)} + \cfrac{1}{K_3 * k_5} * \cfrac{1}{(B)} + \cfrac{1}{K_1 * K_3 * k_5} * \cfrac{1}{(A)*(B)} + \cfrac{1}{k_5}} \qquad (10.14)$$

$$k_{cat} = k_5$$

$$= \frac{k_b * T}{h} * \exp\left[\frac{-\Delta G_5^{\ddagger}}{R*T}\right] \qquad (10.15)$$

The value of the k_{cat}/K_A, or V_{max}/K_A*E_t, as well as that of the k_{cat}/K_B will be a function of magnitudes of all of the energy transitions between the intermediate enzyme form to which the respective substrate binds (as the second substrate) to the transition state of k_5 (equation 10.16). If the release of products has transition states of significant energy, these parameters will be affected by transitions to the latter.

$$\frac{k_{cat}}{K_A} = \cfrac{1}{\cfrac{1}{K_1 * k_5}}$$

$$= \frac{k_b * T}{h} * \cfrac{1}{\exp\left[\cfrac{\Delta G_1^{\circ} + \Delta G_5^{\ddagger}}{R*T}\right]} \qquad (10.16)$$

The value of the K_A as well as that of the K_B, however, are determined by the respective energy differences of the intermediates subtending the respective binding steps (equation 10.17).

$$K_A = \frac{1}{K_1}$$

$$= \frac{h}{k_b * T} * \exp\left[\frac{\Delta G_1^{\circ}}{R * T}\right]$$

(10.17)

10.2. Environmental Effects on Initial Velocity

Although the foregoing relationships are interesting in a theoretical way, it is of more experimental importance to use them to interpret the results of experiments in which the initial velocity is measured under different conditions that affect these energy differences, pH, isotopic substitution, temperature, *etc.* The comparison of two initial velocity measurements under the same conditions except for one is best done by a ratio of the two.

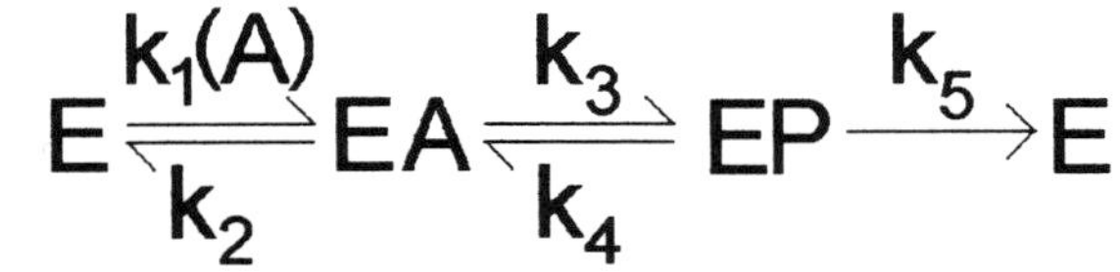

$$E \underset{k_2}{\overset{k_1(A)}{\rightleftharpoons}} EA \underset{k_4}{\overset{k_3}{\rightleftharpoons}} EP \overset{k_5}{\longrightarrow} E$$

Figure 10.6

For the purpose of illustration we use the unireactant chemical model with two intermediates (Figure 10.6). The mathematical model (equation 10.18) for the initial velocity is the same as that given above except that the term for the concentration of substrate has been omitted in the interest of algebraic convenience. It will cancel out in the final equation anyway.

$$\frac{v_i}{E_t} = \frac{1}{\dfrac{1}{k_1} + \dfrac{1}{K_1 * k_3} + \dfrac{1}{K_1 * K_3 * k_5} + \dfrac{1}{k_3} + \dfrac{1}{K_3 * k_5} + \dfrac{1}{k_5}}$$

(10.18)

The ratio of the initial velocity under a set of arbitrarily defined standard conditions, v_0, to that under a set of experimental conditions, v_e, in which one condition has been changed is:

$$\frac{v_o}{v_e} = \frac{\dfrac{1}{k_{1e}} + \dfrac{1}{K_{1e}*k_{3e}} + \dfrac{1}{K_{1e}*K_{3e}*k_{5e}} + \dfrac{1}{k_{3e}} + \dfrac{1}{K_{3e}*k_{5e}} + \dfrac{1}{k_{5e}}}{\dfrac{1}{k_{1o}} + \dfrac{1}{K_{1o}*k_{3o}} + \dfrac{1}{K_{1o}*K_{3o}*k_{5o}} + \dfrac{1}{k_{3o}} + \dfrac{1}{K_{3o}*k_{5o}} + \dfrac{1}{k_{5o}}}$$

(10.19)

In the interest of intuitive and algebraic simplicity the denominator terms can be replaced by a summation (equation 10.20) where $1/(k_0*\prod K_0)$ represents the terms that have one rate constant and can have zero to 3 equilibrium constants (for the model above).

$$\frac{v_o}{v_e} = \frac{\dfrac{1}{k_{1e}} + \dfrac{1}{K_{1e}*k_{3e}} + \dfrac{1}{K_{1e}*K_{3e}*k_{5e}} + \dfrac{1}{k_{3e}} + \dfrac{1}{K_{3e}*k_{5e}} + \dfrac{1}{k_{5e}}}{\displaystyle\sum \dfrac{1}{k_0*\prod K_0}}$$

(10.20)

Equation 10.20 can be further simplified to:

$$\frac{v_o}{v_e} = S*\left[\frac{1}{k_{1e}} + \frac{1}{K_{1e}*k_{3e}} + \frac{1}{K_{1e}*K_{3e}*k_{5e}} + \frac{1}{k_{3e}} + \frac{1}{K_{3e}*k_{5e}} + \frac{1}{k_{5e}}\right]$$

(10.21)

where

$$S = \frac{1}{\displaystyle\sum \dfrac{1}{k_0*\prod K_0}}$$

(10.22)

Multiply each of the terms of the sum in brackets in equation 10.21 by the ratio of the analogous terms under standard conditions, $k_0*\prod K_0/(k_0*\prod K_0)$, (equation 10.23).

$$\frac{v_o}{v_e} = \frac{k_{10}}{k_{1e}}*\frac{1}{k_{10}}*S + \frac{K_{10}*k_{30}}{K_{1e}*k_{3e}}*\frac{1}{K_{10}*k_{30}}*S + \frac{K_{10}*K_{30}*k_{50}}{K_{1e}*K_{3e}*k_{5e}}*\frac{1}{K_{10}*K_{30}*k_{50}}*S +$$

$$\frac{k_{30}}{k_{3e}}*\frac{1}{k_{30}}*S + \frac{K_{30}*k_{50}}{K_{3e}*k_{5e}}*\frac{1}{K_{30}*k_{50}}*S + \frac{k_{50}}{k_{5e}}*\frac{1}{k_{50}}*S$$

(10.23)

The latter equation rearranges to equation 10.24,

$$\frac{v_o}{v_e} = \frac{k_{10}}{k_{1e}} * \Theta_1 + \frac{(K_1 * k_3)_0}{(K_1 * k_3)_e} * \Theta_{13} + \frac{(K_1 * K_3 * k_5)_0}{(K_1 * K_3 * k_5)_e} * \Theta_{15} +$$

$$\frac{k_{30}}{k_{3e}} * \Theta_3 + \frac{(K_3 * k_5)_0}{(K_3 * k_5)_e} * \Theta_{35} + \frac{k_{50}}{k_{5e}} * \Theta_5$$

(10.24)

where:

$$\Theta = \frac{1}{k_o * \prod K_o} * S = \frac{\dfrac{1}{k_o * \prod K_o}}{\dfrac{1}{k_{1o}} + \dfrac{1}{K_{1o} * k_{3o}} + \dfrac{1}{K_{1o} * K_{3o} * k_{5o}} + \dfrac{1}{k_{3o}} + \dfrac{1}{K_{3o} * k_{5o}} + \dfrac{1}{k_{5o}}}$$

(10.25)

For example:

$$\Theta_{13} = \frac{\dfrac{1}{K_{1o} * k_{3o}}}{\dfrac{1}{k_{1o}} + \dfrac{1}{K_{1o} * k_{3o}} + \dfrac{1}{K_{1o} * K_{3o} * k_{5o}} + \dfrac{1}{k_{3o}} + \dfrac{1}{K_{3o} * k_{5o}} + \dfrac{1}{k_{5o}}}$$

(10.26)

The Θ terms represent "determinancy" coefficients and are a measure of the extent to which that particular energy (chemical) transition is rate determining. It follows that:

$$0 \leq \Theta \leq 1.0$$

(10.27)

and

$$\sum \Theta = 1.0$$

(10.28)

However, the mathematical meaning of the determinancy coefficients depends on the model and the parameter derived. For example if the same derivation is applied to the operational parameters, k_{cat}/K_A or k_{cat}, for the chemical model above, a similar mathematical model results (*e.g.* equation 10.29).

$$\frac{\left(\dfrac{k_{cat}}{K_A}\right)_0}{\left(\dfrac{k_{cat}}{K_A}\right)_e} = \frac{k_{10}}{k_{1e}} * \Theta_1 + \frac{\left(K_1 * k_3\right)_0}{\left(K_1 * k_3\right)_e} * \Theta_{13} + \frac{\left(K_1 * K_3 * k_5\right)_0}{\left(K_1 * K_3 * k_5\right)_e} * \Theta_{15} \qquad (10.29)$$

The determinancy coefficients, however, have a somewhat different mathematical meaning in the latter context. For example Θ_{13} in equation 10.30 has a different meaning than that in equation 10.26.

$$\Theta_{13} = \frac{\left(\dfrac{1}{K_1 * k_3}\right)_0}{\dfrac{1}{k_{10}} + \left(\dfrac{1}{K_1 * k_3}\right)_0 + \left(\dfrac{1}{K_1 * K_3 * k_5}\right)_0} \qquad (10.30)$$

Nevertheless, equation 10.27 and equation 10.28 remain valid in the context of the mathematical model for a particular parameter. However, the use of the value for the determinancy coefficient from one parameter, e.g. v_i, in the equation for a different parameter, e.g. k_{cat}/K_A, is invalid even though they may have the same designation.

If the basic energy equations (equation 10.2 and equation 10.3) are substituted into equation 10.24, the ratio of the initial velocities for the chemical model under consideration under two different sets of conditions is expressed in terms of the sum of the differences in the individual energy transitions from each intermediate to every downstream transition state with which it is connected by reversible equilibria times the extent to which that transition is rate limiting (equation 10.31).

$$\frac{v_o}{v_e} = \sum \exp\left[\frac{\Delta G_0^{\ddagger} + \sum \Delta G_0^0 - \left[\Delta G_e^{\ddagger} + \sum \Delta G_e^0\right]}{R * T}\right] * \Theta \qquad (10.31)$$

For example the ratios of the k_{cat}/K_A, under the two different sets of conditions for the unireactant chemical model above would be (equation 10.32):

$$\frac{\left(\dfrac{k_{cat}}{K_A}\right)_0}{\left(\dfrac{k_{cat}}{K_A}\right)_e} = \exp\left[\frac{\Delta G_{10}^{\ddagger} - \Delta G_{1e}^{\ddagger}}{R*T}\right]*\Theta_1 + \exp\left[\frac{\Delta G_{10}^{\circ} + \Delta G_{30}^{\ddagger} - (\Delta G_{1e}^{\circ} + \Delta G_{3e}^{\ddagger})}{R*T}\right]*\Theta_{13} +$$

(10.32)

$$\exp\left[\frac{\Delta G_{10}^{\circ} + \Delta G_{30}^{\circ} + \Delta G_{50}^{\ddagger} - (\Delta G_{1e}^{\circ} + \Delta G_{3e}^{\circ} + \Delta G_{5e}^{\ddagger})}{R*T}\right]*\Theta_{15}$$

The ratio of the values k_{cat} for the unireactant chemical model above would be (equation 10.33):

$$\frac{(k_{cat})_0}{(k_{cat})_e} = \exp\left[\frac{\Delta G_{30}^{\ddagger} - \Delta G_{3e}^{\ddagger}}{R*T}\right]*\Theta_3 +$$

(10.33)

$$\exp\left[\frac{\Delta G_{30}^{\circ} + \Delta G_{50}^{\ddagger} - (\Delta G_{3e}^{\circ} + \Delta G_{5e}^{\ddagger})}{R*T}\right]*\Theta_{35} + \exp\left[\frac{\Delta G_{50}^{\ddagger} - \Delta G_{5e}^{\ddagger}}{R*T}\right]*\Theta_5$$

Therefore, the effect of a change in a single reaction condition on the initial velocity or its operational parameters is expressed as the weighted average of the effect on the individual energy transitions that determine that parameter.

An example of experimental results of this model for environmental effects on a parameter in a two-transition reaction would be an equation (equation 10.34) with two

$$\frac{\left(\dfrac{k_{cat}}{K_A}\right)_0}{\left(\dfrac{k_{cat}}{K_A}\right)_e} = \frac{k_{10}}{k_{1e}} * \Theta_1 + \frac{\left(K_1 * k_3\right)_0}{\left(K_1 * k_3\right)_e} * \Theta_{13}$$

$$\Theta_1 = \frac{\left(\dfrac{1}{k_1}\right)_0}{\left(\dfrac{1}{k_1} + \dfrac{1}{K_1 * k_3}\right)_0}$$

(10.34)

$$\Theta_{13} = \frac{\left(\dfrac{1}{K_1 * k_3}\right)_0}{\left(\dfrac{1}{k_1} + \dfrac{1}{K_1 * k_3}\right)_0}$$

terms. If the change in reaction condition (e.g. isotopic substitution) reduces the rate of one step (k_3) to half its original value with no affect on either the rate of the other step, k_1, or the equilibrium constant, K_1, so that the intrinsic ratio of the rate constants for the affected transition ($K_1 k_3$) is 2.0, the effect on the ratio of the observed parameters, e.g. k_{cat}/K_A, varies from 2.0 to 1.0 (Figure 10.7) depending on the ratio (k_3/k_1) of the rate constants for the affected transition to that of the other transition. Thus, an intrinsic effect (2.0 in this example) is expressed more or less in the measured parameter depending upon the value of the determinancy coefficients. When the affected step is very slow, i.e. $k_3 < k_1$, the measured effect approaches the

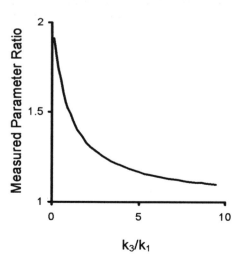

Figure 10.7

intrinsic effect. When the affected step is fast, $i.e.$ $k_3>k_1$, the measured effect approaches 1.0. It is interesting that the measured ratio is about 1.5, when the two rate constants are equal.

10.3. Summary

In this chapter the relationship between initial velocity as well as its operational parameters and the energy transitions associated with the conversion of each intermediate to every downstream transition state with which it is connected by reversible steps has been derived. In addition the conceptual and mathematical model was developed for the effect of changes in the magnitude of these energy transitions on the initial velocity, and its operational parameters. In the following chapters the effects of changes in specific environmental conditions will be related to the initial velocity by virtue of the changes in the energy of specific chemical transitions. In this way the effect of changes in environmental conditions on the kinetic parameters can be at least partially interpreted in terms of the chemical mechanism.

10.4. References

1. Albery, J. and Knowles, J.R. "Evolution of Enzyme Function and the Development of Catalytic Efficiency," *Biochemistry*, 15, 5631-40 (1976).

CHAPTER 11

EFFECTS OF PH

11.1. Introduction

The present chapter will be an exploration of useful kinds of experiments for the measurement of the effects of changes in pH on the initial velocity of an enzyme-catalyzed reaction, of the logic for the interpretation of data in support or rejection of mechanistic hypotheses, and the limitations of both the experiments and the interpretation. These hypotheses are particularly important, when protons are either reactants or products of the reaction and when acid/base catalysis is an issue in the mechanism.

The effects of pH on initial velocity are most frequently determined simply in order to define the optimum pH at which to assay the enzyme activity. For these experiments it is generally satisfactory to determine the initial velocity under a standard set of conditions except for the variable pH. However, the present chapter deals with the determination and interpretation of the effects of pH in order to eliminate mechanistic hypotheses about the enzyme. The operational objective of the latter experiments is to determine the effect of pH on the steady-state operational parameters k_{cat}, (V_{max}) and k_{cat}/K_M (V_{max}/K_M).

The effects of pH are caused by protonation or deprotonation of one of the participant molecules in one or more of the reaction steps. Most commonly, when an essential basic group is protonated or an essential acidic group is deprotonated, the reaction is inhibited completely, but more subtle effects are also observed. It is also possible that the substrate is protonated or deprotonated. Be sensitive to this possibility and determine whether the pK_a's of the substrate could explain the changes observed. Most commonly the protonation or deprotonation of the some enzyme form is the cause of pH effects.

Ideally attempts will be made to determine the identity of both the enzyme form and the chemical group whose protonation or deprotonation causes the change in rate. However, the number possible chemical models exceeds the number of distinguishable mathematical models. In addition several other uncertainties in the interpretation of results limit the success in attainment of this ideal.

11.2. Experiments and Data

Rigorous studies of the effects of pH on the initial velocity parameters are rather labor intensive. The basic experimental objective is the measurement of the initial velocity as a function of substrate concentration in the reaction mixture and calculation of the value of k_{cat} (V_{max}) and k_{cat}/K_M (V_{max}/K_M). These measurements are repeated at a number of different pH values. Therefore, there are generally at least five measurements at each pH. With multireactant enzymes is may be necessary to measure the initial velocity as a function of the concentration of more than one substrate at each pH value. Although their usefulness will be explained, the number of experiments multiplies considerably, particularly if inhibition and other phenomena are studied as a function of pH. Investigators are referred to previous descriptions of the experiments for additional details [1].

11.2.1. THE EXPERIMENTS

The initial velocity and its operational parameters are determined at a variety of pH values with the same kinds of experiments described earlier. Different buffer solutions are utilized to control the pH of the reaction mixtures at different pH values. The buffer in the reaction mixtures are most conveniently made from two stock solutions, one of the acidic and one of the basic forms of the buffer. Make the solutions of equal ionic strength by the addition of an inert salt (usually sodium chloride) to the stock solution of the buffer form of lower ionic strength, although the specific effect, if any, of the salt on the initial velocity should be determined. When constituting the reaction mixture, mix the two buffer forms in different proportions, from approximately 8:2 to 2:8, so that the total buffer concentration is the same in each reaction. The exact proportion is not critical, since the pH should be measured before and/or after each reaction anyway.

Since it is important to perform experiments over as wide a range of pH as possible, it is generally impossible to cover the complete pH-range to be studied with a single buffer. Therefore, it is necessary to use multiple buffers, one or more which may exert specific effects on the initial velocity. There are several ways in which the problem can be circumvented, or at least defined. Buffer mixtures have been defined [2] that contain multiple buffering species that cover a wide pH range and have approximately the same ionic strength throughout the range. Therefore, all of the buffering species, at least in some form, will be present at all times, although specific effects of one or more buffer species will not be apparent.

Alternatively different single buffers can be used for segments of the pH range to be studied, if certain precautions are taken. Try to use chemically similar buffers for all segments, for example the zwitterionic amine buffers cover a wide pH range and are chemically similar. At the intersection of two segments of the pH range perform at least one measurement with each of both buffers at the same pH for the adjacent segments in order to produce an overlap of the segments. In this way the possibility of specific buffer effects on the initial velocity can usually be eliminated or at least defined. Although it is better to select buffers that have no specific buffer effects at the overlapping pH, as a last resort some investigators have either normalized one of the noncoincident values to the other or normalized all of the parameter values to some central value, which is in that portion of the pH profile in which there is minimum change with pH.

11.2.2. PRIMARY DATA MANIPULATION

The methods and programs described in Chapter 3 can be used to calculate the values of V_{max} (or k_{cat}) and V_{max}/K_M (or k_{cat}/K_M), which can then be modeled as a function of pH according to the models below in an attempt to determine the best model and the apparent values of the pK_a's of the acids and/or bases involved.

For reasons that will become apparent the data and models are discussed in the context of the log_{10} of the ratio of the parameter at some arbitrary standard pH, usuallly in a region of the curve in which the change in the parameter value with pH is minimal, divided by the value at the experimental pH. Graphic plots are generally the negative log_{10} of this same ratio. It should be realized that the graphic plot will have the same shape as a plot of log_{10} of the experimental parameter itself.

11.3 Models and Hypotheses

The modeling objectives include the identification of the intermediate enzyme forms or steps that are sensitive to pH and of the base or acid on that intermediate. Although unfortunately neither of these objectives can be attained in all cases with certainty, they can be approximated in most cases and attained in some. Furthermore, in favorable cases information can be generated with respect to the extent to which the pH-dependent transitions are rate determining. However, generally the results of additional experiments are required to attain these objectives with a degree of certainty. The immediate objective of the discussion that follows is to establish a framework within which the investigator can develop models for individual enzymes.

In most proteins, including enzymes, a number of chemical groups will accept or donate a proton, function as bases or acids respectively. However, those groups become important in the present context only if the catalytic activity of the enzyme is different as a result of either protonation or deprotonation. Therefore, there are two mechanistic requirements for a pH effect. There must be an acidic or basic group on the enzyme and that group must have some effect on the catalytic activity.

The pH models will be developed in the context of the general model for environmental effects developed in the previous chapter, *e.g.* equation 10.24. It is useful to discuss the effect of pH on a single-step irreversible reaction of an enzyme form and then to discuss the effect of pH on an equilibrium constant of a reversible, single-step reaction. Finally the effect of pH on a product of one rate constant times one or more equilibrium constants and then the models for the effect on initial velocity and its operational parameters will be addressed. The derivation of mathematical models of detailed chemical models for pH effects requires considerable algebraic labor and many of them cannot be distinguished from each other. Therefore, the objective here will be to show how the mathematical models can be developed, show the various patterns that can be distinguished, and the limitations in the information produced.

11.3.1. ONE-STEP, IRREVERSIBLE MODEL

A model for the effect of pH on the rate constant of a single-step, irreversible reaction will be illustrated for an active basic form of an enzyme intermediate. The pH will have an effect because it determines the distribution of the active enzyme (E_1) in the total of E_1 and HE_1. The pH-dependent, rate constant, k_e, is expressed as a function of the pH-independent rate constant, k, the pH and the pK_a of the basic group involved.

The chemical model (Figure 11.1) with a completely inactive acid form of the enzyme intermediate results in a mathematical model in which the distribution of E_1 and HE_1 (E_{1tot}) is

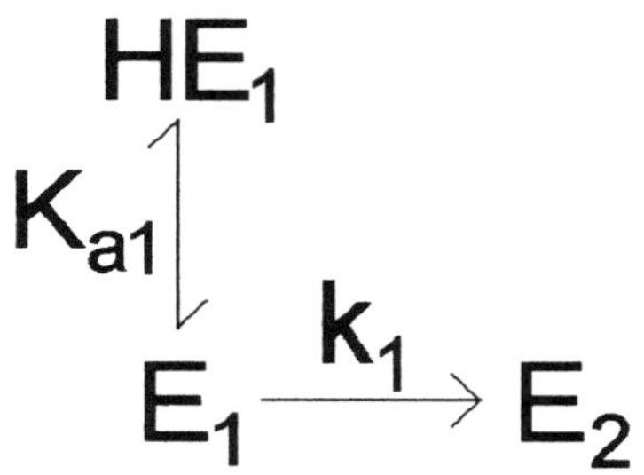

$$HE_1$$
$$K_{a1} \uparrow \downarrow$$
$$E_1 \xrightarrow{\ k_1\ } E_2$$

Figure 11.1

$$f_{EI} = \frac{E_1}{E_{1tot}} = \frac{1}{1 + \dfrac{(H^+)}{K_{a1}}}$$

$$v = k_e * (E_{1tot}) = \frac{k_1}{1 + \dfrac{(H^+)}{K_{a1}}} * (E_{1tot}) \qquad (11.1)$$

$$k_e = \frac{k_1}{1 + \dfrac{(H^+)}{K_{a1}}}$$

treated as a rapid-equilibrium segment as described previously (Chapter 5). The velocity of the reaction (equation 11.1) is the pH -dependent rate constant times the total of both enzyme forms, E_1 plus HE_1. Therefore the pH-dependent rate constant equals the pH-independent rate constant times a pH function.

A plot (Figure 11.2) of the negative $\log_{10}$ of the ratio of the pH-dependent rate constant at some standard pH, k_0, divided by its value at the experimental pH, k_e, vs the pH will approach a horizontal asymptote at high pH, where $(H^+) \ll K_{a1}$, (equation 11.2) and an asymptote with slope=1.0 at low pH, where $(H^+) \gg K_{a1}$. Two asymptotes will intersect where the pH equals the pK_a of the essential base. This graphic pattern will be called the essential base pattern.

The chemical model (Figure 11.3) for an essential acidic enzyme form results in a mathematical model derived in the same way but with a different pH function (equation 11.3).

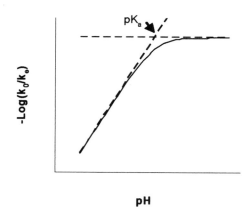

Figure 11.2

High pH: $Log_{10}(k_e) = Log_{10}(k_1)$

Low pH: $Log_{10}(k_e) = Log_{10}(k_1) - pK_{a1} + pH$ (11.2)

Intercept: $pH = pK_{a1}$

$$HE_1 \xrightarrow{\ k_1\ } HE_2$$
$$K_{a1} \downarrow$$
$$E_1$$

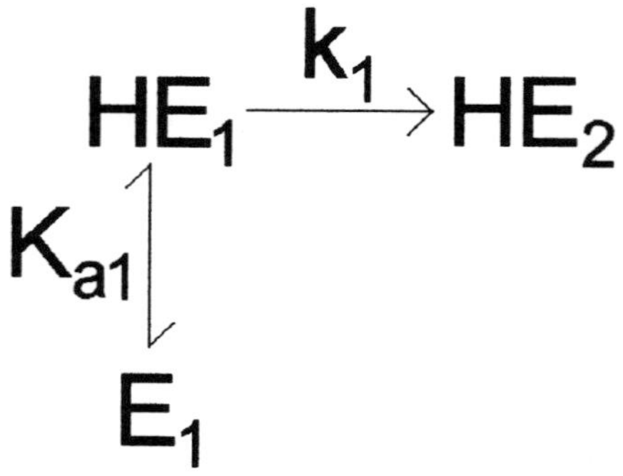

Figure 11.3

$$k_e = \frac{k_1}{1 + \dfrac{K_{a1}}{(H^+)}}$$ (11.3)

A plot (Figure 11.4) of the negative $\log_{10}$ of the ratio of the pH-dependent rate constant at some standard pH divided by the value at the experimental pH *vs* the pH will approach a horizontal asymptote at low pH (equation 11.4) and an asymptote with slope=-1.0 at high pH. Two asymptotes will intersect where the pH equals the pK_a of the acid (equation 11.4). This pattern will be called the essential acid pattern.

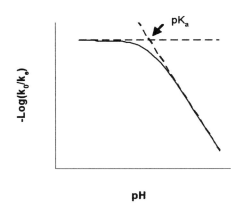

pKa
-Log(k0/ke)
pH

Figure 11.4

$$\begin{aligned}
&\textit{Low pH:} && Log_{10}(k_e)=Log_{10}(k_1)\\
&\textit{High pH:} && Log_{10}(k_e)=Log_{10}(k_1)+pK_{a1}-pH \qquad (11.4)\\
&\textit{Intercept:} && pH=pK_{a1}
\end{aligned}$$

The chemical model (Figure 11.5) for an active ampholyte form of the enzyme intermediate, in which both the relative acid and basic forms are inactive is derived in the same way from the distribution of active enzyme (HE_1) in the total of that enzyme form (E_1+HE_1+$H2E_1$). It results in a mathematical model in which the pH function is a kind of combination of the two pH functions above (equation 11.5).

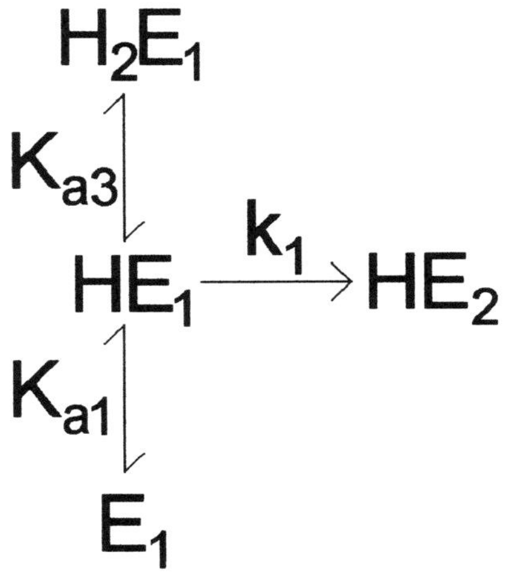

H2E1
Ka3
HE1
k1
HE2
Ka1
E1

Figure 11.5

$$k_e = \frac{k_1}{1 + \frac{K_{a1}}{(H^+)} + \frac{(H^+)}{K_{a3}}}$$ (11.5)

A plot of the negative $\log_{10}$ of the ratio of the pH-dependent rate constant at some standard pH divided by the value at the experimental pH vs the pH (Figure 11.6) will approach an asymptote with slope $=1.0$ at low pH and another with slope $=-1.0$ at high pH with a horizontal segment between. The asymptotes will intersect where the pH equals the pK_a values depending upon how close the values of the acid and base pK_a's are to each other. The further apart they are the closer

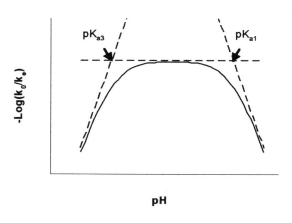

Figure 11.6

the graphical approximation to the pK_a values. They should be at least two pH-units apart for a reasonably accurate graphic approximation. However, the fit of equation 11.5 to the data will result in the best estimates for the values of the pK_a's in any case.

It can be shown that the slopes of the non-horizontal asymptotes in all of the models above reflect the number of protons involved in the association or dissociation of the base or acid respectively. For example a slope of 2.0 reflects an active base or bases that accept two protons. It is also possible to distinguish more than one base or acid with different pK_a values, if the values of their respective pK_a's are not too close together.

If the acid or base permutation of the most active enzyme form is only less active (rate constant k_1') and not completely inactive, as in the previous examples, the observed, pH-dependent rate constant is the sum of the intrinsic rate constants of the pH-permuted enzyme forms each multiplied by the appropriate pH function (e.g. equation 11.6).

$$k_e = \frac{k_1}{1 + \frac{K_{a1}}{(H^+)}} + \frac{k_1'}{1 + \frac{(H^+)}{K_{a1}}}$$ (11.6)

A plot (Figure 11.7) of the $\log_{10}$ of the pH-dependent rate constant *vs* the pH will be a wave function with horizontal asymptotes at the pH extremes.

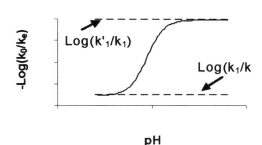

Figure 11.7

11.3.2. ONE-STEP REVERSIBLE MODEL.

Although the effects of pH on the equilibrium constant of a single-step reaction are somewhat more complex, when first considered, there are a couple of redundancies that simplify the models somewhat, because of the properties of equilibrium constants themselves.

The chemical model (*e.g.* Figure 11.8) in which both the reactant and product enzyme forms have an inactive, pH-permutation results in a mathematical model (equation 11.7) in which the pH-dependent equilibrium constant is related to the pH-independent equilibrium constant by the ratio of two pH functions. Of course, if the values of the pK_a's are the same, the two will cancel and there will be no effect of pH on the equilibrium constant.

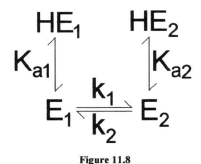

Figure 11.8

$$K_{1e} = K_1 * \frac{1 + \dfrac{(H^+)}{K_{a2}}}{1 + \dfrac{(H^+)}{K_{a1}}} \tag{11.7}$$

A chemical model in which both the reactant and product are active acids or bases, but the pK_a has changed substantially (Figure 11.8) results in a mathematical model (equation 11.7) in which pK_{a1} is different from pK_{a2}. The graphical plot of the apparent equilibrium constant will be a wave similar to that in Figure 11.7, in which the two horizontal asymptotes will intersect the asymptote with a slope = 1.0 where the pH equals each of the two pK_a values.

The chemical model (*e.g.* Figure 11.9) in which only the reactant enzyme form has an inactive, pH-permutation results in a mathematical model (equation 11.8) in which the pH-dependent equilibrium constant is related to the pH-independent equilibrium constant in a similar manner to that in which the respective rate constants were related above (equation 11.1, Figure 11.2) .

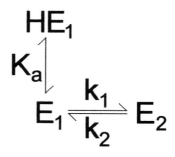

Figure 11.9

$$K_{1e} = \frac{K_1}{1 + \frac{(H^+)}{K_a}}$$

(11.8)

11.3.3.THE pH EFFECTS ON THE PRODUCT OF RATE AND EQUILIBRIUM CONSTANTS

The next step in the development of models for pH effects is the compilation of the models above into the individual terms in the generalized model for environmental effects, of which each term is a product of a rate constant and zero to several equilibrium constants. After examination of the effects of pH on the individual terms of the generalized equation, further compilation of the latter into the effects on the sum of such terms that constitute the generalized model will be discussed. (See equation 10.24, Chapter 10.)

The compilation of the pH-effects on rate and equilibrium constants into the effects on the product of a rate constant and none to several equilibrium constants results in simpler models than expected, but the excess of numbers of possible chemical models over the number of distinguishable mathematical models limits the certainty of the interpretation. In addition there are too many possible models to derive them all here. Therefore, the objectives here will be to demonstrate some general principles for the effects with a representative sample of the possible models and to demonstrate an approach to model development.

If those terms that contain only a rate constant and no equilibrium constant are pH-dependent or if the term has both rate and equilibrium constants but only the rate constant is affected by pH, they will conform to one of the models for a rate constant described above. Furthermore, since the pertinent parameters are actually ratios of the value at some standard pH divided by the value at the experimental pH, the intrinsic rate constants cancel out (*e.g.* equation 11.9)

$$\frac{k_0}{k_e} = 1 + \frac{K_a}{(H^+)}$$

(11.9)

However, in those terms that contain one or more equilibrium constants the effects of pH can be somewhat more complicated. If all of the enzyme forms represented in the term have the same type of essential dissociable groups (Figure 11.10), any pH function incorporated in the pH-dependent rate constant, *e.g.* k_5,

Figure 11.10

will cancel out with the pH function in the reverse reaction contained in the adjacent, upstream equilibrium constant (equation 11.10). In addition the pH function associated with the pH-dependent forward rate constant included in any equilibrium constant will cancel out with the pH function in that of the reverse, pH-dependent rate constant in the equilibrium constant of the next upstream step.

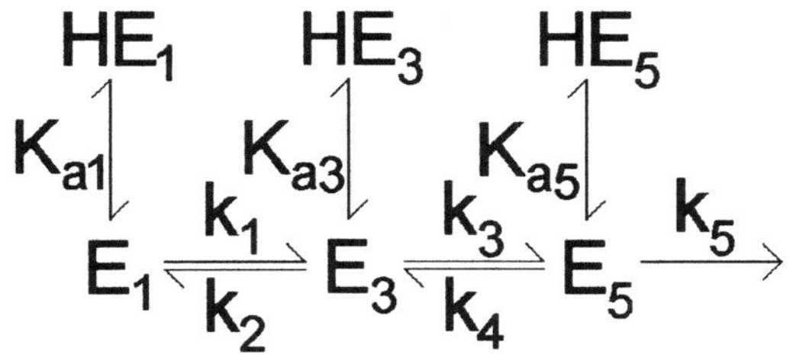

$$(K_1 * K_3 * k_5)_e = K_1 * \frac{1 + \frac{(H^+)}{K_{a3}}}{1 + \frac{(H^+)}{K_{a1}}} * K_3 * \frac{1 + \frac{(H^+)}{K_{a5}}}{1 + \frac{(H^+)}{K_{a3}}} * k_5 * \frac{1}{1 + \frac{(H^+)}{K_{a5}}}$$

$$= K_1 * K_3 * k_5 * \frac{1}{1 + \frac{(H^+)}{K_{a1}}}$$

(11.10)

Therefore, the pH function expressed in the multi-step transitions is that of first enzyme form in the transition as long as there are no additional complications and the pK_a's of the essential groups associated each intermediate enzyme form are the same for the forward and reverse reactions of each step (equation 11.10). Furthermore, since the pertinent experimental parameters are actually ratios of the value at some standard pH divided by the value at the experimental pH (equation 10.24), the intrinsic rate constants cancel out (*e.g.* equation 11.11).

$$\frac{(K_1 * K_3 * k_5)_0}{(K_1 * K_3 * k_5)_e} = 1 + \frac{(H^+)}{K_{al}}$$

(11.11)

This simplification is not so surprising, when it is remembered that the rate of a given transition depends only on the energy difference between the initial intermediate enzyme form and the downstream transition state and not the energy of any of the intermediates between. Therefore, it is logical that the pH-permutations of the intermediate enzyme forms would have an effect on the overall transition only in special circumstances.

If the dissociable groups are no longer essential for one or more of the steps represented by equilibrium constants in the term in question (Figure 11.11), the mathematical model for the overall term is unaffected. In the latter case the equilibrium constant K_3 equals K'_3, then $K_{a3} = K_{a5}$ and the two corresponding pH functions cancel out in equation 11.12.

Figure 11.11

Therefore, the pH effects on the transition in Figure 11.11 are the same as those for the model in Figure 11.10.

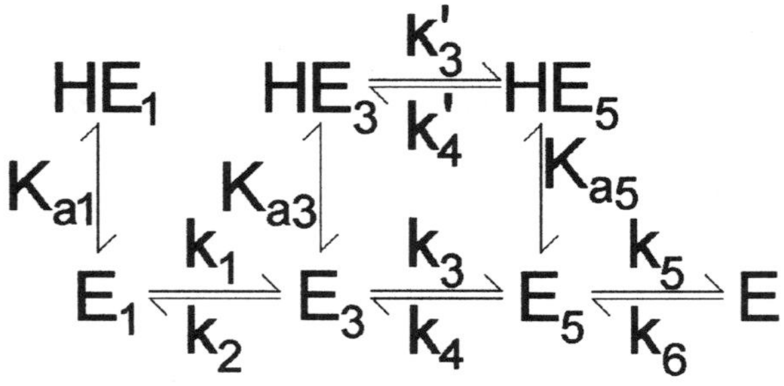

$$(K_1 * K_3 * k_5)_e = K_1 * \frac{1 + \dfrac{(H^+)}{K_{a3}}}{1 + \dfrac{(H^+)}{K_{al}}} * K_3 * k_5 * \frac{1}{1 + \dfrac{(H^+)}{K_{a5}}}$$

$$= K_1 * K_3 * k_5 * \frac{1}{1 + \dfrac{(H^+)}{K_{al}}}$$

(11.12)

Furthermore it can be shown that this holds true whether the pH-permuted form, e.g. HE_3, reacts in the forward direction at the same rate or a different rate as the non permuted form.

However, if the dissociable groups on the last enzyme form in the transition are no longer essential (Figure 11.12), the mathematical model for the overall term is a ratio of the pH functions for the reverse reaction for the last enzyme form in the transition and the forward reaction for the first enzyme form in the transition (equation 11.13) which results in a wave in the graphical plot. Since in this model there is the uptake of one proton in one of the branches of the reaction pathway, one of the transition states has a different protonation status from that of the first enzyme form in the transition. The effects of proton uptake is discussed more fully below.

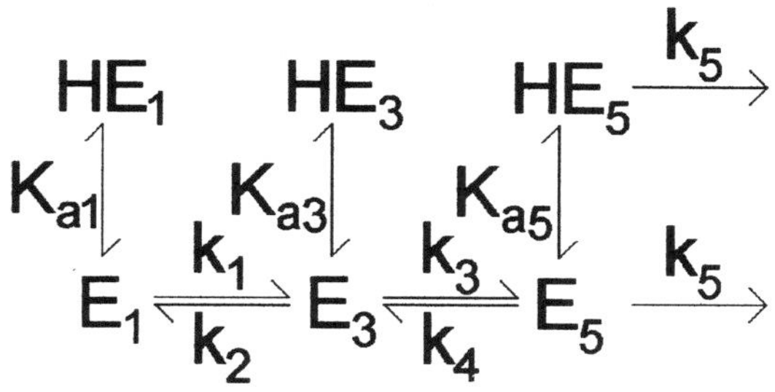

The major assumption in the models above was that the

$$HE_1 \qquad HE_3 \qquad HE_5 \xrightarrow{k_5}$$

$$E_1 \underset{k_2}{\overset{k_1}{\rightleftarrows}} E_3 \underset{k_4}{\overset{k_3}{\rightleftarrows}} E_5 \xrightarrow{k_5}$$

Figure 11.12

$$(K_1 * K_3 * k_5)_e = K_1 * \frac{\left[1 + \dfrac{K_{a3}}{(H^+)}\right]}{\left[1 + \dfrac{K_{a1}}{(H^+)}\right]} * K_3 * \frac{\left[1 + \dfrac{K_{a5}}{(H^+)}\right]}{\left[1 + \dfrac{K_{a3}}{(H^+)}\right]} * k_5$$

$$= K_1 * K_3 * k_5 * \frac{\left[1 + \dfrac{K_{a5}}{(H^+)}\right]}{\left[1 + \dfrac{K_{a1}}{(H^+)}\right]}$$

(11.13)

essential pK$_a$'s of the intermediate enzyme forms are the same in both the forward and reverse directions. The only way in which the essential pK$_a$'s for the forward and reverse steps of a given intermediate to be different is for the essential chemical groups themselves to be different for the forward and reverse steps. However, the possible dimensions of differences are too numerous to describe them all, but any of the graphical patterns, and models, described above may be produced. In addition the patterns may have graphical notches or bumps in the essential base, the essential acid or the active ampholyte pattern. Other models that also produce these latter patterns are described below.

Essential proton release or uptake for the reaction may take place involving one of the intermediate enzyme forms in a transition. It will be manifest on the mathematical model for the pH effects on that transition. The form of the mathematical model will depend upon the effects of pH on the initial and final enzyme forms in the transition. For example if none of the

$$HE_3 \underset{k_4}{\overset{k_3}{\rightleftharpoons}} E_5 \xrightarrow{k_5}$$

$$\Big\downarrow K_{a3}$$

$$E_1 \underset{k_2}{\overset{k_1}{\rightleftharpoons}} E_3$$

Figure 11.13

other enzyme forms have essential dissociable groups (Figure 11.13), proton uptake in one of the intermediate enzyme forms causes a proton to appear as a substrate or product in the transition (equation 11.14). If the transition were the dominant one for the experimental parameter of interest, the resulting graphic pattern will be indistinguishable from that of an essential acid, because at high pH the pH-independent terms in the overall expression will become dominant and constant.

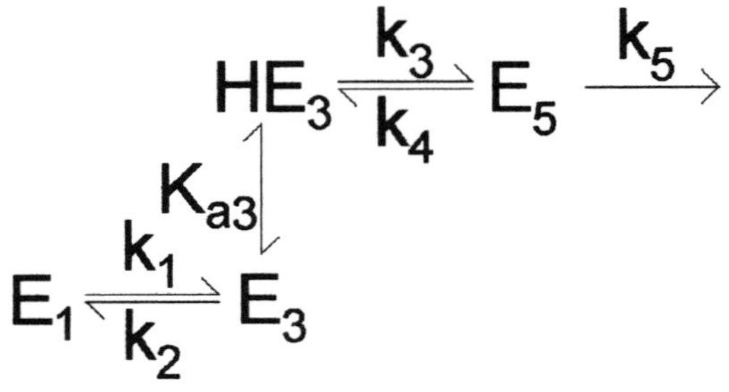

$$(K_1 * K_3 * k_5)_e = K_1 * \left[1 + \frac{(H^+)}{K_{a3}}\right] * K_3 * \frac{1}{\left[1 + \frac{K_{a3}}{(H^+)}\right]} * k_5$$

$$= K_1 * K_3 * k_5 * \frac{(H^+)}{K_{a3}}$$

(11.14)

However, if the first enzyme form has an essential base in the pH range of the experiments, proton uptake in one of the intermediate enzyme forms (Figure 11.14) results in a mathematical model that contains the pH function for the essential basic dissociable group (equation 11.15). If this were the dominant transition for the steady-state parameter determined experimentally, the pattern would be indistinguishable from that for a simple essential acidic group.

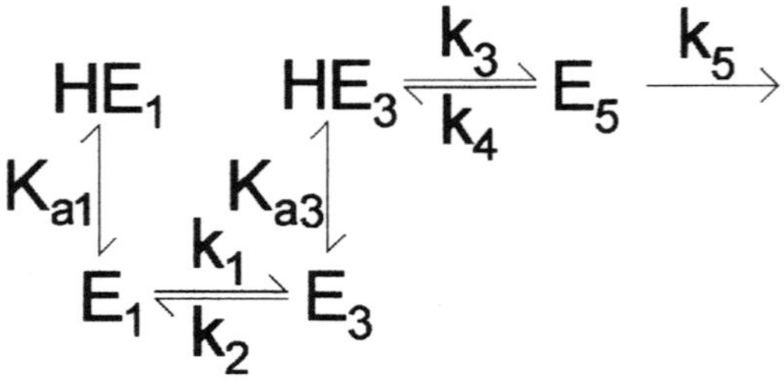

$$HE_1 \qquad HE_3 \underset{k_4}{\overset{k_3}{\rightleftharpoons}} E_5 \xrightarrow{k_5}$$

$$\Big\downarrow K_{a1} \qquad \Big\downarrow K_{a3}$$

$$E_1 \underset{k_2}{\overset{k_1}{\rightleftharpoons}} E_3$$

Figure 11.14

$$(K_1 * K_3 * k_5)_e = K_1 * \cfrac{1}{\left[1 + \cfrac{(H^+)}{K_{al}}\right]} * K_3 * k_5 * \cfrac{(H^+)}{K_{a3}}$$

$$= K_1 * K_3 * k_5 * \cfrac{K_{al}}{K_{a3}} * \cfrac{1}{\left[1 + \cfrac{K_{al}}{(H^+)}\right]}$$

(11.15)

The other patterns resulting from models for proton uptake or release are also indistinguishable from those resulting from mathematical models that do not include proton uptake or release. Therefore, hypotheses of proton uptake or release is impossible to distinguish from other possibilities on the basis of these experiments alone.

11.3.4. THE pH EFFECTS ON INITIAL VELOCITY PARAMETERS

The models for the effect of pH on the ratio of the initial velocities of a unireactant chemical model, *e.g.* with an essential base in every enzyme form (Figure 11.15), are developed from the generalized model for environmental effects, (equation 10.24 reproduced here as equation 11.16).

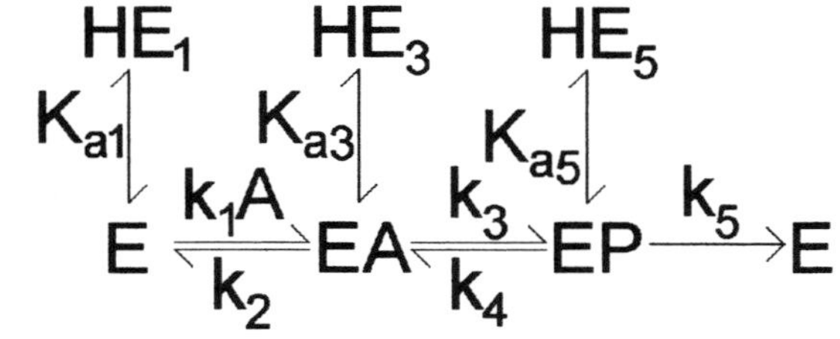

The mathematical model

Figure 11.15

$$\frac{v_o}{v_e} = \frac{k_{10}}{k_{1e}} * \Theta_1 + \frac{K_{10} * k_{30}}{K_{1e} * k_{3e}} * \Theta_{13} + \frac{K_{10} * K_{30} * k_{50}}{K_{1e} * K_{3e} * k_{5e}} * \Theta_{15} +$$

$$\frac{k_{30}}{k_{3e}} * \Theta_3 + \frac{K_{30} * k_{50}}{K_{3e} * k_{5e}} * \Theta_{35} + \frac{k_{50}}{k_{5e}} * \Theta_5$$

(11.16)

for pH (equation 11.17) is a sum of the pH effects on each of the transitions involved in the reaction each ratio multiplied by a determinancy coefficient that expresses the extent to which that transition is rate determining.

Therefore, the total effect on the initial velocity is beset by a new range of complications and uncertainties. Furthermore, the fact that the patterns of experimental

$$\frac{v_o}{v_e} = \left[1 + \frac{H^+}{K_{a1}}\right] * \Theta_1 + * \left[1 + \frac{H^+}{K_{a1}}\right] * \Theta_{13} + \left[1 + \frac{H^+}{K_{a1}}\right] * \Theta_{15} +$$

$$\left[1 + \frac{H^+}{K_{a3}}\right] * \Theta_3 + \left[1 + \frac{H^+}{K_{a3}}\right] * \Theta_{35} + \left[1 + \frac{H^+}{K_{a5}}\right] * \Theta_5$$

(11.17)

data are generally fairly simple and, therefore, frequently indistinguishable from each other render the modeling effort of questionable value. However, a few general principles are useful even if only to define the limitations of this approach.

The graphical plot of the negative $\log_{10}$ of the ratio of initial velocities *vs* the pH may be some combination of the patterns of an essential acid group, an essential basic group and one or more waves. However, most experimental results have the fairly simple pattern of the curves shown above and the data can be fit to those pH models. A common reason for the prevalence of relatively simple experimental patterns is suggested below.
If the pH-dependent transition is rate determining, *i.e.* has a value of Θ close to 1.0, the apparent pK_a will be the intrinsic pK_a, that of some intermediate enzyme form in that transition. However, if the pH-dependent transition is fast, has a low value of Θ, its rate must be made more slow before it will be expressed in the initial velocity. Therefore, in the latter case the apparent pK_a of an essential acidic group will be higher than the intrinsic pK_a and the apparent pK_a of an essential basic group will be lower than the intrinsic pK_a. The pK_a's are pushed to the pH extremes, when the transition is not rate determining. For example the ratio of the initial velocities for a chemical model with an essential basic group on one intermediate will have two kinds of terms (equation 11.18), those containing a pH function and those containing no pH function.

$$\frac{v_o}{v_e} = \left[1 + \frac{H^+}{K_{a1}}\right] * \Theta_1 + \Theta_3$$

$$= \left[1 + \frac{H^+}{K_{a1}}\right] * \Theta_1 + 1 - \Theta_1$$

$$= 1 + \Theta_1 * \frac{H^+}{K_{a1}}$$

(11.18)

At high pH the graphic plot will approach the horizontal and at low pH it will have approach a slope of 1.0. The apparent pK_a, the intersection of the asymptotes, will then be less than the intrinsic pK_a, if the value of Θ_1 is less than 1.0 (equation 11.19). Thus the apparent pK_a of essential basic groups will be lower than the intrinsic pK_a. Conversely that of essential acidic groups will be higher than the respective intrinsic pK_a. It seems likely that some of the pH-sensitive transitions that are not rate determining will be out of the range of the experiments. This latter likelihood contributes to the relative simplicity of the patterns of experimental data.

$$High\ pH: \qquad -Log_{10}\left[\frac{v_0}{v_e}\right]=0$$

$$Low\ pH: \quad -Log_{10}\left[\frac{v_0}{v_e}\right]=-Log_{10}(H^+)-Log_{10}\left[\frac{\Theta_1}{K_{a1}}\right] \qquad \text{(11.19)}$$

$$Intercept: \qquad pH=pK+Log_{10}\Theta_1$$

In addition if there are multiple pH-sensitive transitions, more than one of which are rate-determining, the observed pK_a will be a weighted average of them. For example the ratio of the initial velocities for a chemical model with an essential basic group on two intermediate enzyme forms, each with a different pK_a (equation 11.20), will have at least two kinds of terms, one for the pH function of each essential group.

$$\frac{v_o}{v_e}=\left[1+\frac{(H^+)}{K_{a1}}\right]*\Theta_1+\left[1+\frac{(H^+)}{K_{a3}}\right]*\Theta_3$$

$$=\left[1+\frac{(H^+)}{K_{a1}}\right]*\Theta_1+\left[1+\frac{(H^+)}{K_{a3}}\right]*\left[1-\Theta_1\right] \qquad \text{(11.20)}$$

$$=1+(H^+)*\left[\frac{\Theta_1}{K_{a1}}+\frac{\Theta_3}{K_{a3}}\right]$$

Thus the apparent pK_a for an essential basic group will be an average of the two weighted by the determinancy coefficients (equation 11.21).

Furthermore, more than one rate determining, pH-sensitive transition may have different pH patterns. In the latter case the experimental results may be combination of

$$High\ pH: \qquad -Log_{10}\left[\frac{v_0}{v_e}\right]=0$$

$$Low\ pH: \quad -Log_{10}\left[\frac{v_0}{v_e}\right]=-Log_{10}(H^+)+Log_{10}\left[\frac{1}{\dfrac{\Theta_1}{K_{a1}}+\dfrac{\Theta_3}{K_{a3}}}\right] \qquad (11.21)$$

$$Intercept: \qquad pH=-Log_{10}\left[\frac{1}{\dfrac{\Theta_1}{K_{a1}}+\dfrac{\Theta_3}{K_{a3}}}\right]$$

patterns. For example simulation of the combination of an essential base pattern with a wave pattern, which have equal determinancy coefficients (0.5), produces a notch in the essential base pattern (Figure 11.16). Some adjustment of the pK_a values in the simulation can produce a hump in the essential base pattern as well.

Ordinarily the pH effects are calculated for k_{cat} and the k_{cat}/K_M because the number and kinds of transitions involved in each of these parameters are more restricted. The pH effects on k_{cat}/K_M should reflect acidic or basic groups on the intermediate enzyme form to which

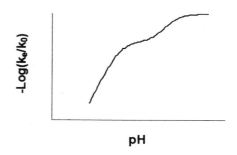

Figure 11.16

that substrate binds. However, if that enzyme form is bounded upstream by one or more reversible steps that do not involve the binding of other substrates the pH effects may reflect essential acidic or basic groups on the reactant enzyme form in the latter steps also. In addition if that intermediate enzyme form is not affected by pH in the range measured, the effect may be due to a different intermediate. Moreover, net proton uptake or release in one or more of these transitions may complicate the model. Nevertheless in the simple model the pH effects on this parameter should reflect the intrinsic pK_a of the enzyme form to which the substrate binds. A strategic approach to circumstantial confirmation of the simple model is presented below.

The pH effects on k_{cat} reflect acid or basic groups on all of the enzyme forms that do not bind substrate, *e.g.* catalytic, and/or product-release steps.

11.3.4. IDENTIFICATION OF CHEMICAL GROUPS

Although apparent values of pK_a's can be obtained with acceptable precision from the curve fits of data from pH experiments to the most appropriate of the simple mathematical models 11.1, 11.3, 11.5, 11.6, and 11.7, the assignment of specific intermediate enzyme forms and known chemical groups to them is severely limited by several uncertainties. First are the uncertainties associated with the models themselves and the identification of the critical enzyme form. Second the value of the intrinsic pK_a of the essential acid or base may be different from that of the apparent pK_a, unless the transitions from the intermediate enzyme form for which that acid or base is essential are rate determining. Third, the intrinsic value of the pK_a of a group on an intermediate enzyme form may be quite different from that of the same chemical group on a free amino acid or even on a small peptide, because of the different dielectric constant and hydrogen bonding opportunities associated with the microenvironment in the enzyme protein.

Although they are all rather laborious, three methods have been employed in attempts to circumvent the uncertainties and provide additional evidence for the identity of the intermediate enzyme forms and the chemical groups involved: temperature perturbation, solvent perturbation and judicious selection of the steady-state parameters upon which the pH effects are determined.

The pH experiments have been performed at more than one temperature (usually two temperatures) in order to calculate the enthalpy, ΔH, of the ionization [3]. The pK_a values of uncharged acids, $e.g.$ carboxylic acids, change rather little with temperature (small ΔH), whereas the pK_a values of cationic acids, $e.g.$ protonated amines, change significantly with temperature (larger ΔH). However, other important factors may change with temperature also. If the enthalpy associated with a transition from the intermediate in question is significant that transition may become more or less rate determining with the temperature change and thereby change the apparent pK_a value. Furthermore, the microenvironment of the group in question may change with temperature.

In the second approach the pH experiments have been repeated in the presence of different solvents of lower dielectric constant, usually mixtures of water and either dioxane, ethyl alcohol, or acetone [4]. A solvent of lower dielectric constant will lower the dissociation constant, raise the pK_a, of a neutral acid, whereas it will raise the dissociation constant, lower the pK_a, of a cationic acid. However, additions of organic solvents will also affect the hydrogen-bonding properties of the solvent mixture [5] and there remains some uncertainty about the effect of the solvent on the microenvironment in which the essential acid or base is found.

Evidence to diminish all of the uncertainties in the assignment of apparent pK_a values to specific enzyme forms and chemical groups can be generated by the determination of pH effects on equilibrium parameters instead of rate parameters. Thus, the pH effect is determined on the value the dissociation constant, K_i, of a product inhibitor or an analog inhibitor or the K_M values of substrates in rapid equilibrium with the enzyme. Alternatively the pH effect can be determined in direct binding experiments. Comparison of the apparent

pK_a for the latter parameters with that for the k_{cat}/K_M for the same chemical species as a substrate may give an indication of the extent to which the transitions of the enzyme form that bind the substrate are rate limiting. The identity of these two measured pK_a-values provides circumstantial evidence for a simple model, in which there are no contiguous reversible steps upstream from the binding step and no stoichiometric protons. Therefore, the apparent pK_a must be the same as the intrinsic pK_a, and it must reflect a group on the enzyme form to which the substrate binds. However, it must be stressed that such evidence is circumstantial.

11.4. Summary

The systematic investigation of the pH effects on initial velocity and its parameters is a labor-intensive undertaking and the interpretation is limited by an array of limitations and uncertainties. For example an enzyme with two substrates would require approximately twenty-five initial-velocity measurements at each pH in order to construct the curves for k_{cat}, k_{cat}/K_A and k_{cat}/K_B. However, some things can be learned from the investigations.

The fact that there are frequently more chemical models than distinguishable mathematical models makes it impossible to eliminate all but one of the former. The fact that the observed pK_a value is frequently different from that of the chemical group responsible for the pH-effect makes the identification of these groups uncertain.

Nevertheless certain patterns of pH effects on kinetic parameters support a limited number of hypotheses. For example an apparent essential base can be distinguished from an essential acid. In addition the number of protons involved in the pH-permutation of the essential groups can be determined. Furthermore if the pH effects on k_{cat}/K_M reflect the pH perturbations of the enzyme form to which that substrate binds, the same pH effect should be seen on the K_i for that substrate. In the latter case the apparent pK_a should be the intrinsic pK_a.

In distinction to the pH effects on kinetic parameters those on equilibrium parameters, K_i values for analog or product inhibitors, are determined by the actual pK_a of the essential groups in their individual microenvironments. These values are useful in two lines of further investigation. First, further investigation of the effects of temperature and solvent on these pK_a values along with a comparison of the data with the existing tables of pK_a-ranges and temperature effects for known amino acid residues may permit tentative identification of the groups involved. Second, agreement of these pK_a values with those for the k_{cat}/K_M for the corresponding substrate is confirmatory evidence that the latter values reflect those of the essential groups on the enzyme form to which the substrate binds. Nonagreement indicates that some complicating factor influences the latter: a reversible step immediately upstream from the enzyme form to which the substrate binds, stoichiometric protons, or a change in essential groups of some intermediate in the upstream and downstream directions.

Finally, knowledge of the pH effects allows the investigator to change the rate determining transitions by carrying out experiments at a nonoptimum pH as part of an investigation of other environmental effects.

11.5. References

1. Cleland, W.W. "The Use of pH Studies to Determine Chemical Mechanisms of Enzyme-Catalyzed Reactions," *Methods Enzymol.* 87, 390-405 (1982).

2. Ellis, K.J. and Morrison, J.F. "Buffers of Constant Ionic Strength for Studying pH-Dependent Processes," *Methods Enzymol.* 87, 405-26 (1982).

3. Cleland, W.W. "Determining the Chemical Mechanisms of Enzyme-Catalyzed Reactions by Kinetic Studies," *Adv. Enzymol Relat. Areas Metab.* 45, 323 (1977).

4. Cleland, W.W. "Determining the Chemical Mechanisms of Enzyme-Catalyzed Reactions by Kinetic Studies," *Adv. Enzymol Relat. Areas Metab.* 45, 320-48 (1977).

5.Grace, S. and Dunaway-Mariano, D. "Examination of the Solvent Perturbation Technique as a Method to Identify Enzyme Catalytic Groups," *Biochemistry,* 22, 4238-47, (1983).

CHAPTER 12

EFFECTS OF ISOTOPIC SUBSTITUTION

12.1. Introduction

In this chapter a discussion is presented of the effects of isotopic substitution on initial velocity and its operational parameters. Although isotopic substitution, particularly with radioactive isotopes, has long been used to detect and measure the rate of enzymatic reactions, it also alters the rate and equilibrium of chemical reactions, if the bonding to the isotopic atom is changed during the course of the reaction. Therefore, the systematic investigation of these effects can reveal the importance and nature of bond changes to the reaction rate and equilibrium.

The objective of the chapter is for the reader to know the kinds of experiments that are most useful, the format of a data produced in each as well as the limitations of the data produced in each. An additional objective is for the reader to understand the models of isotope effects in order to plan useful experiments, and to interpret data from these experiments as well as those in the current literature.

There is a significant and useful literature on this subject. Although the modeling context in most is somewhat different from that presented here, the conclusions are the same. The reader is referred to some of the more extensive and specialized presentations, particularly for the experimental details [1],[2].

12.1.1. SCOPE AND ORGANIZATION

The general subject of isotope effects in chemical reactions is a very broad one. At one extreme the general subject overlaps the fields of archeology and ecology, particularly of plants, whereas at another extreme it overlaps the fields of quantum mechanics and reaction rate theory. Although the latter fields particularly are important for a completely rigorous understanding of the effects of isotopic substitution on enzymatic reactions, they will be covered here only in sufficient detail to permit initiation of the logic that leads to a working understanding. The reader will be referred to more detailed sources for a discussion of the more basic aspects.

The most useful kinds of experiments as well as their technical requirements and limitations will be described first. Then the models for the interpretation of data will be developed, and the associated limitations will be described. The limitations will suggest additional experiments to eliminate them.

12.1.2.KINDS OF ISOTOPE EFFECTS

Generally three types of isotope effects will be discussed here: primary, secondary and solvent isotope effects.

Primary isotope effects are associated with isotopic substitution of an atom subtending a covalent bond broken or made during the course of the reaction.

Secondary isotope effects are associated with isotopic substitution of an atom subtending a covalent bond which may be changed but not broken or made during the course of the reaction. These bonds are frequently spatially close to the bonds that are broken or made.

Solvent isotope effects are changes in the rate or equilibrium of the reaction due to substitution of deuterium for hydrogen in the solvent water in which the reaction takes place. Of course all of the exchangeable (noncovalent) protons are also substituted with deuterium to an extent depending on the atom-per-cent excess of the isotope.

The investigation of primary and secondary isotope effects is significantly difficult and laborious, but the interpretation of the data is more likely to result in useful information about the reaction mechanism. The experimental investigation of solvent isotope effects is significantly easier conceptually. However, it can also be quite laborious, if it is carried out rigorously. In addition the interpretation of data from the latter experiments is somewhat limited because of the large number of possible processes and protons involved.

12.1.3. SYMBOLISM AND NOTATION

In spite of efforts to the contrary the symbolism of isotope effects on enzymatic reactions remains somewhat inconsistent.

For the isotopes of hydrogen $^D V/K$, or $^D k_{cat}/K_M$, and $^D V$, or $^D k_{cat}$, denote the deuterium (^2H) isotope effects on V_{max}/K_M or k_{cat}/K_M and on V_{max}, and k_{cat}, respectively. Whereas $^T V/K$ denotes the tritium (^3H) isotope effects on V_{max}/K_M [3]. However, these symbols are not universally employed and different symbols are used for isotopes of heavy atoms. For example a common designation for isotope effects of ^{13}C is simply $^{12}k/^{13}k$. This designation seems adequate in principal, since these are almost always isotope effects on V_{max}/K_M, but this fact is not apparent from the designation. A more satisfactory and more generally employed designation is $^{13}V_{max}/K_M$. In the remainder of this book the value of a parameter in the presence of the lower-abundance, hearier isotope will be designated by a subscript i or a subscript designating the isotope itself, $e.g.$ k_D for the rate constant with deuterium substitution.

The concise designation of secondary and solvent isotope effects is even more difficult than that of primary isotope effects. Frequently it is the same as that for primary isotope effects with the explicit designation in the text that it refers to secondary or solvent isotope effects. Solvent isotope effects have been designated with a superscript D_2O, but that seems to be beyond the capabilities of most word processors and printers.

12.2. Experiments and Results

There are three general methods for the measurement of isotope effects. However, not all are equally applicable to all three kinds of isotope effects. The methods will be presented first in the context of primary isotope effects and then their application to secondary and solvent isotope effects will be discussed. Finally some methods and precautions for the determination of solvent isotope effects will be presented.

It must be realized that before any measurements of primary or secondary isotope effects can be made, the proper isotopically labeled compounds must be acquired and characterized. Most commonly the compounds must be synthesized and characterized by the investigator, and these tasks can require a very significant amount of ingenuity, time and labor. Since it is necessary for the two substrates to be as identical as possible except for isotopic substitution, it is frequently advisable to synthesize and characterize both the normal-abundance and the isotopically enriched compound in parallel. Although chemical synthesis is usually the first method considered, enzymatic synthesis, frequently with employment of the same enzyme as that under investigation, is frequently more convenient. The acquisition of the appropriate isotopically labeled compounds will not be considered further here.

12.2.1. MEASUREMENT BY DIRECT COMPARISON

The most elementary method for the determination of isotope effects is direct comparison of the values of the initial velocity and its operational parameters, k_{cat} and k_{cat}/K_M, in the presence and absence of isotopically labeled substrate as described in Chapter 2. This method has the advantage that the isotope effects on all of the steady-state parameters can be determined. However, it has the disadvantage that its sensitivity is limited by the precision with which they can be determined. Therefore, since the magnitude of the isotope effect is an inverse function of the atomic mass of the atom under consideration, the isotope effects of carbon, nitrogen and oxygen are generally an order of magnitude smaller than those of hydrogen isotopes (Table 12.1), and the method is limited to the determination of the primary isotope effects of the deuterium isotope of hydrogen, ^2H and the determination of solvent isotope effects. Determination of the primary isotope effects of other atoms and of secondary isotope effects generally require more sensitive techniques.

12.2.2. MEASUREMENT BY COMPETITION

Isotope effects can be measured with the greatest sensitivity by competition. The reaction is conducted in the presence of both isotopic and non-isotopic substrate together, and the ratio of the two isotopes is measured either in the product or in the remaining substrate as a function of the fraction of the total reaction that has transpired. Measurements of the ratio in product are made early in the reaction, $(P_i/P)_t$, and very late in the reaction, $(P_i/P)_\infty$, regarded as infinite time. The isotope effect is calculated on the first order rate constant for the reaction by equation 12.1, in which "f" is the fraction of reaction at the early point, and the subscript i designates the parameter associated with the heavier isotope [4]. If the early point can be made very early, <15%, the formula is significantly simplified (equation 12.2). The reader is referred to the more detailed literature [4] for the applicable analogous formula, when the ratio of isotope is measured in substrate.

$$\frac{k}{k_i} = \frac{\ln[1-f]}{\ln\left[1-f*\left(\frac{(P_i/P)_t}{(P_i/P)_\infty}\right)\right]}$$

(12.1)

$$\frac{k}{k_i} = \frac{(P_i/P)_\infty}{(P_i/P)_t}$$

(12.2)

Since V_{max}/K_M, or the k_{cat}/K_m, for an enzymatic reaction is the first-order rate constant for the reaction, the isotope effect on the first-order rate constant calculated from the formula above will be the isotope effect on the V_{max}/K_M for that substrate.

Because of their low abundance the competition method is the only one that can be used with radioactive isotopes. Although the most intuitively obvious application of this method would be the use of a single radioactive isotope with the measurement of the changing specific activity of either the product or the substrate, the precision of measurement is compromised by the fact that the measurement of radioactivity and mass must be done on separate samples. In order to improve the precision double labeling is used with one isotope in the chemically sensitive position in one population of substrate and the other in a relatively insensitive position of a separate population of the substrate. After the two populations are mixed and the reaction initiated, the ratio can be determined directly on a single sample by double-isotope counting. (See reference [5] for details.)

The extraordinary precision of measurement of the ratio of isotopes by the isotope-ratio mass spectrometer makes competition the ideal method for the determination of isotope effects of the heavy atoms of nitrogen, oxygen and carbon. In fact this method can frequently be used with substrates of normal isotopic abundance. However, since these instruments generally require gaseous samples, some chemistry must be done on the samples to release the isotope in this form, unless the isotope is released as a gas during the reaction, *e.g.* carbon dioxide. Therefore, the investigator must be sensitive to the possibility of isotope effects during sample preparation. The reader is referred to more detailed references for chemical methods of sample preparation and handling [4].

Sample preparation is made easier, or at least more uniform, if one population of substrate molecules is labeled with a heavy atom at some atom remote from a kinetically sensitive one, particularly a position that can be converted conveniently to a gaseous compound. The kinetically sensitive atom is labeled with one isotope in one population that is also labeled remotely, whereas the kinetically sensitive atom contains the other isotope in the another population of the same substrate that is otherwise unlabeled. Then the reaction is carried out in the presence of both populations and the ratio of the isotopes from the remote position can be measured either in the substrate or the product as described above [6].

12.2.3. MEASUREMENT BY EQUILIBRIUM PERTURBATION

For reactions that are significantly reversible and one of whose substrates or products can be measured continuously in real time independently of the other products and reactants, *e.g.* with a spectrophotometer, the technique of equilibrium perturbation will determine the isotope effects with most of the heavy-atoms. The reaction is set up with the reactant and product concentrations as close to equilibrium as possible with one reactant, or one product, labeled at the kinetically sensitive position with the heavy isotope in the highest abundance possible and the opposite product, or reactant, respectively containing the normal isotope. After it is established that the concentration of the reactant or product measured is as stable as possible, the enzyme is added. If there is an isotope effect, the reaction of the unlabeled reactants will proceed at a higher rate than that of the labeled one and there will be a perturbation of the equilibrium, which will come back to a constant concentration of the reactants and products as isotopic equilibrium is approached. The magnitude of the perturbation is proportional to the isotope effect and the isotope effect on V_{max}/K_M can be calculated with the equations derived by Cleland [7]. The equilibrium isotope effect can also be calculated.

12.2.4. MEASUREMENT OF SOLVENT ISOTOPE EFFECTS

Although solvent isotope effects are determined by direct comparison, there are several phenomena that complicate the experiments. The isotope exchange into all of the exchangeable chemical positions should be at equilibrium. In addition there is a solvent isotope effect on the pK_a of buffers as well as of the acidic and basic groups associated with the enzyme and substrate, which necessitates some additional experiments. Furthermore, there is an isotope effect on the electrode of the pH-meter itself (*c.a.* 0.4 pH units), for which the reading can be corrected.

The equilibrium of the isotope exchange is promoted by dissolution of all of the components of the reaction mixture in D_2O solutions. In order to measure a stable isotope effect all of the exchangeable protons in the reaction mixture must ideally be exchanged for deuterons. Although the exchangeable protons in the substrates and other small molecules generally exchange rapidly, some of those in the enzyme protein may be notably slow. In the absence of an independent measurement of complete exchange at least an acceptably stable fraction of exchange is achieved, if the enzyme is stored in D_2O buffer for a number of hours (*e.g.* overnight) before use.

Finally the isotope effect on the pK_a's can be dealt with by determination of the steady-state parameters as a function of pH (p^2H) in isotopic water and comparison with the data in nonisotopic water. The isotope effect on the parameter is the ratio of the parameters at a horizontal segment of the pH profile providing there is a significant such segment (*e.g.* 1-2 pH units). If the segment is attenuated, the maximum values can be obtained from a fit to the pH functions in the previous chapter. If the pH curves are not done or if the values compared are on the rapidly-changing (slope $\geq |1.0|$) segment of the pH curve, it is possible the measured isotope effect will include the effect of isotopic solvent on the pK_a as well as that on the parameter itself.

12.3. The Origin of Isotope Effects

Isotope effects originate in the nature of chemical bonding itself, the rigorous understanding of which requires an understanding of some detailed concepts and logic of quantum mechanics. Therefore, in consideration of succinctness, and the author's background limits, the presentation will be rather intuitive and the reader is referred to more detailed references on the subject [8], [9], [10], [11].[1]

Isotopic substitution results in a number of energy differences in a chemical compound or bond. Although the energy of all chemical bonds is partitioned into a number of different energy levels under generally ambient conditions, the energy of the bonds subtended by heavier isotopes partition into somewhat lower energy levels than those with the lighter isotopes. This has a couple of consequences.

First in an equilibrium situation the heavier isotope will partition into the compound in which the associated bond energy is lower, *i.e.* the tighter bond, and complete isotopic substitution will produce a change in the equilibrium of reactants and products that have different bond energies. Therefore, there are equilibrium isotope effects, due to the difference in the associated energy change, when the heavy isotope goes from reactants to products, from that, when the lighter isotope goes from reactants to products (the difference

---

[1]The absence of quantum-mechanical tunneling is one of the assumptions in the development here of the theory of isotope effects. Although such tunneling has been hypothesized to explain certain isotope effects in enzymatic reactions, this subject will not be covered in the present discussion.

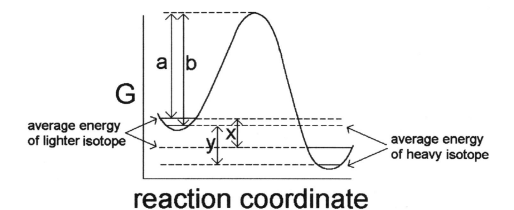

Figure 12.1

between x and y in Figure 12.1). In addition if the average bond energy in the substrates and the products is different, there will be a difference in the distribution of isotopes between the two. Therefore, in the presence of isotope there will be a slightly different equilibrium, an equilibrium isotope effect, and the heavier isotope will be in greater abundance in, partition into, the compound with the lower bond energy.

Second at the point of bond breaking at which all of the original bond is broken and no new bond is formed both isotopes have the same energy or at least the energy levels are very close together (Figure 12.1). Therefore, the energy requirement for breaking bonds subtended by the heavier isotope is greater than that for the lighter one (the difference between a and b in Figure 12.1), and the reaction proceeds more slowly with the heavier isotope. If the bond in question is not completely broken in the transition state, the energy levels are at least closer together and the energy difference, *i.e.* isotope effect, will still be significant but lower in magnitude.

The magnitude of a primary isotope effect on the rate of a reaction is inversely proportional to the amount of chemical bonding to the isotopic atom in the transition state of the rate-determining transition, whereas that of an equilibrium isotope effect depends on the difference in average bonding energy in the substrates and products. The average maximum magnitude of primary isotope effects has been estimated from quantum mechanics (Table 12.1). For more specific values the extent to which heavier isotopes partition into chemical bonds is an indication of the relative magnitude of the possible isotope effect, when that compound is the substrate. Tables of the partitioning of heavier isotopes into various types of chemical bonds compared to some standard compound, φ values, are available [12].

Table 12.1: Approximate average maximum
values of primary kinetic isotope effects by the
common heavy isotopes.

Isotope	k/k_i
^2H	6.87
^3H	16.04
^{13}C	1.054
^{15}N	1.044
^{18}O	1.068

Changes in bond hybridization will also results in isotope effects, and the heavier isotope, *e.g.* deuterium, will partition into the sp^3 hybridization of saturated carbon better than into the sp^2 or the sp hybridization of unsaturated compounds. These latter phenomena are the origin of secondary isotope effects.

An isotope effect less than 1.0 is an inverse isotope effect. Inverse primary covalent isotope effects are due to equilibrium isotope effects. In addition secondary isotope effects on equilibrium or rate may be inverse. Furthermore, because of the complex origin of solvent isotope effects (discussed below), the latter are more commonly inverse isotope effects than are primary covalent isotope effects.

The initial velocity of an enzymatic reaction, as well as its operational parameters, is determined by numerous transitions any or all of which can be partially or completely rate-determining (Chapter 10). An isotope effect on initial velocity can be due to an intrinsic isotope effect on a rate, and effect on an equilibrium or both. In addition, the magnitude of the measured isotope effect will also depend upon the degree to which those transitions that are affected by the isotopic substitution are rate determining.

12.4. Models and Data Interpretation

Models for primary covalent isotope effects will be discussed first, then those for secondary covalent isotope effects and finally those for solvent isotope effects. The principal objective of the interpretation is the determination of the value of the intrinsic isotope effect on rate. In addition there are other mechanistic features, of possible greater interest, that can be elucidated.

12.4.1. PRIMARY ISOTOPE EFFECTS

The discussion of primary and secondary isotope effects will be restricted to those effects on k_{cat}/K_M (or V_{max}/K_M), since this is the effect that is measured in most cases and since the models for this effect are conceptually easier to understand. The most commonly used models for the interpretation of measured isotope effects are in terms of the intrinsic isotope effects on specific reaction steps and the commitment factors, forward and reverse, *e.g.* [13]. However, the context for the models presented here is developed from that for environmental effects developed in the previous chapter, *e.g.* equation 10.29, since the latter context is somewhat easier to understand intuitively and is of somewhat more general application. Therefore, a reasonable model for the isotope effects on k_{cat}/K_A for the same steady-state model (unireactant with two intermediates) is equation 12.3, where the subscript i designates the parameter in the presence of the heavier isotope.

$$\frac{\left[\frac{k_{cat}}{K_A}\right]}{\left[\frac{k_{cat}}{K_A}\right]_i} = \left[\frac{k_1}{k_{1i}}\right] * \Theta_1 + \left[\frac{K_1 * k_3}{K_{1i} * k_{3i}}\right] * \Theta_{13} + \left[\frac{K_1 * K_3 * k_5}{K_{1i} * K_{3i} * k_{5i}}\right] * \Theta_{15} \tag{12.3}$$

A model for the isotope effects on k_{cat}/K_M in other chemical models can be developed by very similar procedures. Nevertheless even in any model all of the transitions will start with substrate in the ground-state energy levels.

There will be three kinds of terms in the mathematical model (equation 12.4): the sum of all the terms with an intrinsic isotope effect on the rate constant for the transition, the sum of all the terms with an intrinsic isotope effect only on an equilibrium constant in

$$\frac{\left[\frac{k_{cat}}{K_A}\right]}{\left[\frac{k_{cat}}{K_A}\right]_i} = \sum \left[\frac{k}{k_i}\right] * \Theta_k + \sum \left[\frac{K}{K_i}\right] * \Theta_K + \sum \Theta \tag{12.4}$$

the transition and the sum of all the terms with no isotope effect.

Intrinsic, Primary Isotope Effects

A convenient situation occurs when the measured isotope effect is at or near the maximum possible for that isotope (Table 12.1), when the determinancy coefficient for the collection of terms that contain the rate constant for the bond breaking is very nearly 1.0. This situation is most likely to be seen in enzymes whose initial velocity fits best to a rapid-equilibrium model, since at least the binding of substrate is not rate determining and those transitions whose transition state is the interconversion of the central, complexes are more likely to be rate determining. However, even in the latter chemical models it is possible that transitions that involve product release will be rate determining [14].

Attempts have been made to achieve the observation of a maximum isotope effect by the use of inferior substrates or pH values for the determination. Thus $^{D}V/K_{malate}$ for dehydrogenation and decaroboxylation of malate by malic enzyme was 1.47 with NADP as the coenzyme but 3.0 with thio-NADP [15].

In those cases in which the measured isotope effect is significantly less than the maximum expected the challenge is to calculate the value of the intrinsic isotope effect. If the significant intrinsic rate isotope effect is due to a single transition state and if the significant equilibrium isotope effect is due to a single equilibrium in the reaction, each of the first two sums in equation 12.4 becomes a single term (equation 12.5).

$$\frac{\left[\dfrac{k_{cat}}{K_A}\right]}{\left[\dfrac{k_{cat}}{K_A}\right]_i} = \left[\frac{k}{k_i}\right]*\Theta_k + \left[\frac{K}{K_i}\right]*\Theta_K + \sum \Theta \qquad (12.5)$$

Although the results with transition-state analogs has supported the hypothesis of bond strain concerted with the binding of substrate, attempts to demonstrate isotope effects in steps other than the catalytic step, *e.g.* a binding step, resulted in no measurable effect. For example the deuterium isotope effect in the binding of NADH to liver alcohol dehydrogenase [16] was 1.0. In any case equation 12.5 has five unknowns: the equilibrium isotope effect (K/K_i) with its determinancy coefficient (Θ_K); the rate isotope effect (k/k_i) with its determinancy coefficient and the determinancy coefficient for the remaining steps.

However, one unknown in equation 12.5 can be eliminated by the estimation of the equilibrium isotope effect by experimental measurement or calculation [17], [7] of the equilibrium isotope effect for the overall reaction. If the estimate of the equilibrium isotope effect is accepted, there are now only four unknowns. Furthermore an additional equation with the same unknowns is provided by the fact that the sum of all the determinancy coefficients is 1.0 (equation 12.6). Thus there are now two equations and four unknowns.

$$\Theta_k + \Theta_K + \sum \Theta = 1.0 \qquad (12.6)$$

Several approaches have been taken toward the elimination, or partial elimination, of the discrepancy between the number of equations and the number of unknowns. Most have involved measurement of the isotope effects of additional isotopes that affect the same transition state.

In those cases in which the effects of isotopes of hydrogen are investigated the Swain-Schaad relationship [18] will provide an additional equation for the intrinsic isotope effect of tritium in terms of the intrinsic isotope effect of deuterium (equation 12.7). The limitations of this relationship are discussed elsewhere, but are not likely to be exceeded

$$\frac{k}{k_T} = \left[\frac{k}{k_D}\right]^{1.44}$$

$$\frac{K}{K_T} = \left[\frac{K}{K_D}\right]^{1.44}$$

(12.7)

in enzymatic experiments. The use of this relationship (equation 12.8) results in four unknowns and three equations in the system to be solved.

$$\frac{\left[\dfrac{k_{cat}}{K_A}\right]}{\left[\dfrac{k_{cat}}{K_A}\right]_D} = \left[\frac{k}{k_D}\right] * \Theta_k + \left[\frac{K}{K_D}\right] * \Theta_K + \Theta$$

$$\frac{\left[\dfrac{k_{cat}}{K_A}\right]}{\left[\dfrac{k_{cat}}{K_A}\right]_T} = \left[\frac{k}{k_D}\right]^{1.44} * \Theta_k + \left[\frac{K}{K_D}\right]^{1.44} * \Theta_K + \Theta$$

(12.8)

$$1.0 = \Theta_k + \Theta_K + \Theta$$

However, the investigator can put limits on the intrinsic isotope effect by use of the limits for the determinancy coefficients, i.e. $0 \leq \Theta \leq 1.0$. This approach was described in a somewhat more complicated theoretical context by Northrop [19]. For example Schimmerlik et al [20] estimated the limits of 5-8 for the deuterium isotope effect on malic enzyme.

An alternative approach requires the measurement of the effect of isotopic substitution at the opposite end of the bond from the original isotope under investigation. For example if the original investigation is of the effect of isotopic substitution of hydrogen in a carbon-hydrogen bond, the effect of carbon isotope in that position could be measured. The reaction would be conducted with two populations of substrate, one with both ^{13}C and

^{2}H and the other with ^{12}C and ^{1}H. Although this contributes an additional equation it also adds another unknown, the intrinsic isotope effect of carbon. However, the isotope effect of both deuterium and carbon simultaneously can be determined to contribute a second additional equation with no additional unknowns. The latter mathematical model will contain a term for the intrinsic effect of deuterium substitution multiplied by the intrinsic effect of a carbon isotope (equation 12.9).

$$\frac{\left[\dfrac{k_{cat}}{K_A}\right]}{\left[\dfrac{k_{cat}}{K_A}\right]_{D,13}} = \left[\frac{k}{k_D}\right]*\left[\frac{k}{k_{13}}\right]*\Theta_k + \left[\frac{K}{K_D}\right]*\left[\frac{K}{K_{13}}\right]*\Theta_K + \Theta \tag{12.9}$$

This last equation results in a system of five equations in five unknowns, which theoretically can be solved, even if the solution must be done numerically.

An alternative approach was taken by Hermes et al [21] with glucose-6-phosphate dehydrogenase by measurement of the deuterium isotope effect on the V/K for glucose-6-phosphate, the carbon-13 isotope effect on the V/K for glucose-6-phosphate and the combination isotope effect. They also determined the effect of a deuterium in the C-4 position of NADP on the V/K for glucose-6-phosphate by direct measurement, a secondary isotope effect, and the combination of the latter effect and the ^{13}C effect on the V/K for glucose-6-phosphate by competition. Although this results in six equations in nine unknowns, three of the unknowns are equilibrium isotope effects that can be estimated independently. Therefore, a solution is possible. However, a limitation arises from the relative imprecision associated wit the determination of the secondary isotope effect by direct measurement.

Catalytic Sequence

With enzymes that catalyze the breaking of more than one major bond the study of multiple isotope effects has been very useful in elucidation of the order of the bond breaking processes. The effect of isotopic substitution on V_{max}/K_M, or k_{cat}/K_M, at one of the sessile bonds is compared with and without isotopic substitution at the other bond. For example in oxidative decarboxylation reactions (*i.e.* malic enzyme) the ^{13}C-isotope effects are compared with and without complete deuterium substitution at the oxidized position [15]. The mathematical model for the isotope effect is different from that described above (equation 12.9), because in the present experiments all of the substrate in any one determination contains either deuterium, ^{2}H, or hydrogen, ^{1}H, in the oxidized position. The model is developed in detail in Appendix 12.1, Section 9.6.

If the two processes are concerted, the ^{13}C-isotope effect will be augmented in the presence of deuterium, because the deuterium will increase the determinancy coefficients for both the transitions with the ^{13}C-rate effect and those with the ^{13}C-equilibrium effect; the deuterium will make the transitions more rate determining.

If the dehydrogenation precedes the decarboxylation, the ^{13}C-isotope effect will be diminished, because the deuterium isotope effect on the rate of dehydrogenation renders more rate determining some transitions that do not involve the ^{13}C-effect, and thus diminish the determinancy coefficient for the ^{13}C isotope effect. Although the ^{13}C-effect on rate will be augmented by the equilibrium isotope effect of deuterium, the latter is rather modest in magnitude.

If the decarboxylation precedes the dehydrogenation, the ^{13}C isotope effect on the rate of the decarboxylation will be unaffected by the deuterium. Although the ^{13}C-effect on equilibrium will be augmented by the deuterium effect on rate, the former cannot contribute more than its rather modest intrinsic value to the overall isotope effect. The investigators determined that substitution of the mobile hydrogen atom with deuterium resulted in a decreased ^{13}C-isotope effect and concluded that the reaction proceeded with dehydrogenation and decarboxylation in separate steps in that order.

Once the sequence of the processes is determined both of the intrinsic isotope effects can be calculated.

12.4.2. SECONDARY COVALENT ISOTOPE EFFECTS

Isotopic substitution of an atom, usually close (α or β) to the reaction center or connected to it with an unsaturated system, but of an atom bound to the same other atoms in the products as in the reactants results in secondary isotope effects, if the bond hybridization to the substituted atom changes in the course of the reaction. They are generally quite small; α-hydrogen isotope effects are generally 1.02-1.4. In addition to their usefulness in the determination of intrinsic, primary isotope effects described above, they have been quite useful in the demonstration of changes in bond hybridization in the course of enzymatic reactions.

For example an enzyme from *Bordatella pertussis* toxin (pertussis toxin) catalyzes the transfer of ADP-ribose from NAD to several proteins, but catalyzes the hydrolysis of nicotinamide from the NAD in the absence of a protein acceptor. In the latter activity a secondary isotope effect of ^{3}H at the 2'-position of the ribose whose glycosidic bond is hydrolyzed, supports the hypothesis that delocalization of electrons between the 1' and 2' carbons takes place in the transition state [22].

12.4.3. SOLVENT ISOTOPE EFFECTS

The interpretation of solvent isotope effects is even more complex than that of covalent isotope effects and the reader should consult some of the excellent detailed treatises for rigorous theory on this subject. The present objective is to identify the dimensions and extent of complexity, describe the limits of certainty of the possible model elimination and estimate the labor involved in their investigation.

Models for solvent isotope effects on a single rate or equilibrium (equation 12.10) are expressed by:

$$\frac{k_{H_2O}}{k_{D_2O}} = \frac{\prod_i^v \varphi^R}{\prod_j^v \varphi^T}$$

$$\frac{K_{H_2O}}{K_{D_2O}} = \frac{\prod_i^v \varphi^R}{\prod_j^v \varphi^P}$$

(12.10)

where φ_j^R, φ_j^T and φ_i^P are the fractionation factors of deuterium into the reactants, the hypothetical transition state and the products respectively compared to water. Tables of these fractionation factors [12] reflect the fact that the higher fractionation factors have tighter binding of the proton than water. The range from j to v and from i to v reflect the numbers of exchangeable protons in the reactants and transition state or products respectively that change in the reaction. Because of the conventions in reporting the fractionation factors the solvent isotope effects in most of the references are expressed as the rate constant in D_2O divided by that in H_2O, which is the inverse of that for other isotope effects. Since there is a significant probability that one or more protons will have stronger bonds in the product or even the transition state than in the reactants, the determinations of inverse solvent isotope effects, less than 1.0 in the convention above, is possible. For example a proton in a sulfhydryl, S-H (φ=0.4), in the reactants might be replaced by one in an hydroxyl, O-H (φ=1.0), in the product.

As an example, the mathematical model (equation 12.11) for the solvent

$$\frac{\left[\dfrac{k_{cat}}{K_A}\right]_{H_2O}}{\left[\dfrac{k_{cat}}{K_A}\right]_{D_2O}} = \left[\frac{\prod_i^v \varphi^R}{\prod_j^v \varphi^T}\right]_1 *\Theta_1 + \left[\frac{\prod_i^v \varphi^R}{\prod_j^v \varphi^T}\right]_{13} *\Theta_{13} + \left[\frac{\prod_i^v \varphi^R}{\prod_j^v \varphi^T}\right]_{15} *\Theta_{15}$$

(12.11)

isotope effects on the k_{cat}/K_A for the chemical model of a unireactant enzymatic reaction (i.e. equation 10.29) is the sum of the effects for each of the transitions each multiplied by its determinancy coefficient. The $\Pi\varphi^R$ is the product of the fractionation factors for all of the exchangeable protons in the free substrate and the enzyme in the absence of that substrate whose bonding is changed in the transition state and the $\Pi\varphi^T$ is the analogous term for the transition state for all of the protons whose bonds are changed from that in the free enzyme and the free substrate. Unfortunately the models for solvent isotope effects are associated with an additional dimension of variability than are covalent isotope effects, due to the multiplicity of exchangeable protons in the enzyme and the substrate as well as that due to the multiplicity of transitions that determine the k_{cat}/K_A of this enzymatic reaction. The mathematical models for other enzymatic chemical models and other steady-state parameters will have similar complications.

In addition some of the assumptions that limited the number of models for covalent isotope effects are no longer reasonable with solvent isotope effects. For example the assumption that no transition in the enzymatic reaction will contain a significant isotope effect, not equal to 1.0, in more than one step is no longer reasonable. Any combination of steps may have both rate and equilibrium isotope effects. Therefore, few of the chemical models for solvent isotope effects can be distinguished from each other with experimental data, and the investigator frequently is obliged to do a certain amount of hand waving to interpret data from the experiments.

However, certain situations and experiments, in which the rate-determining chemical process can be identified, can produce useful data, and some hypothetical models for the rate-determining transition can be supported with data from solvent isotope experiments in cases of prior identification of the rate-determining transitions, *e.g.* by studies of covalent isotope effects. In addition useful, if not portentous, insights into the mechanism can be obtained by determination of the solvent isotope effects on identifiable chemical processes such as binding constants of products and analog inhibitors. However, the rather considerable amount of labor involved may not be considered justifiable for the import of the insights gained.

In those cases in which the principal chemical process can be identified the proton inventory method and theory may eliminate some models for the number and kinds of protons involved in the process. The method requires the determination of the solvent isotope effect in the presence of various atom-per-cent of deuterium in the solvent. The mathematical model (equation 12.12, derivation in [23]) expresses the rate constant in D_2O/H_2O mixtures of various atom-per-cent deuterium, k_n, as a function of the rate constant in H_2O, k_0, the atom-fraction D_2O, n, and the fractionation factors, φ^R and φ^T for all of the involved protons.

$$k_n = k_0 * \frac{\prod_i^\nu \left[1 - n + n * \varphi^T\right]}{\prod_j^\nu \left[1 - n + n * \varphi^R\right]} \tag{12.12}$$

Chemical models with different numbers of involved protons result in mathematical models with different numbers of terms in the products, $\prod$, and slightly different shapes to the graphical plot of k_n *vs* n. Some of these models may be distinguishable from each other. For example if the approximation is made that the isotope effect (*i.e.* bonding changes) associated with each of the protons is the same [24] and results from a ratio of the fractionation factors of 0.35, the predicted plot of the observed isotope effect, k_n, *vs* the atom fraction deuterium is different for different numbers of involved protons (Figure 12.2) due to the mathematical model. However, because of the experimental error (*e.g.* 5% error

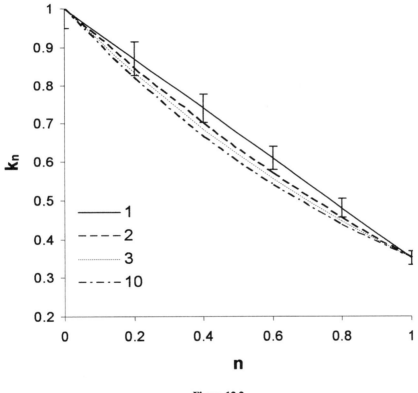

Figure 12.2

bars for the single proton curve in Figure 12.2), in practice it is usually possible to distinguish only rather distant models from each other. In addition it must be remembered that the elimination of a model is more rigorous that the acceptance of a model.

12.5. Chapter Summary

The investigation of isotope effects can accomplish much to confirm or limit the hypotheses about the chemical changes and their relative rates in an enzymatic reaction.   Although the syntheses of reactants and the experimental determination of covalent isotope effects are demanding, the enzymatic models are sufficiently rigorous to permit, in theory, the determination of the relative rates of the bond-breaking transitions and the intrinsic isotope effect in those transitions. Because of their greater magnitude the investigation of primary isotope effects is more rewarding in this objective than that of secondary isotope effects. However, the investigation of the smaller secondary isotope effects can uncover bonding changes more remote from the reacting atoms and severely limit the hypotheses about the reaction mechanism.

Although the determination of solvent isotope effects does not require the chemical or enzymatic synthesis of substrates, the models for their interpretation contain so many dimensions of variability that there is a trade-off between rigor and degree of detail. Therefore, rigorous interpretation is frequently so general as to be meaningless. However, in cases in which one or more of the dimensions of variability can be restricted, either experimentally or by the mechanism itself, the number and/or kinds of protons involved can be determined in proton inventory experiments.

12.5. References

1. Cook, P.F. (Ed.) *Enzyme Mechanisms from Isotope Effects*, CRC Press, Boca Raton, (1991).

2. Cleland, W.W., O'Leary, M.H. and Northrop, D.B. (Eds.) *Isotope Effects on Enayme-Catalyzed Reactions,* University Park Press, Baltimore (1977).

3. Northrop, D.B. "Determining the Absolute Magnitude of Hydrogen Isotope Effects," in W.W. Clelend, M.H. O'Leary and D.B. Northrop (eds.) *Isotope Effects on Enzyme-Catalyzed Reactions,* University Park Press, Baltimore (1977) pp. 122-52.

4. O'Leary, M.H. "Determination of Heavy-Atom Isotope Effects on Enzyme-Catalyzed Reactions," *Methods Enzymol,* 64, 83-104 (1980).

5. Parkin, D.W. "Methods for the Determination of Competitive and Noncompetitive Kinetic Isotope Effects," in P.F. Cook (ed.) *Enzyme Mechanisms from Isotope Effects,* CRC Press, Boca Raton, (1991) pp. 269-290.

6. Kiick, D.M. "Use of the Stable Isotope Remote Label Technique to Determine Isotope Effects for Enzyme-Catalyzed Reactions," in P.F. Cook (ed.) *Enzyme*

Mechanisms from Isotope Effects, CRC Press, Boca Raton, (1991) pp.313-29.

7. Cleland, W.W. "Measurement of Isotope Effects by the Equilibrium Perturbation Technique," *Methods Enzymol.* 64, 104-25 (1980).

8. Melander, L. *Isotope Effects on Reactions Rates,* Ronald Press, New York (1960).

9. Buddenbaum, W.E. and Shiner, V.J., Jr. "Computation of isotope Effects on Equilibria and Rates," in W.W. Cleland, M.H. O'Leary and D.B. Northrop (eds.) *Isotope Effects on Enzyme-Catalyzed Reactions,* University Park Press, Baltimore (1977) pp 1-36.

10. Sühnel, J. and Schowen, R.L. "Theoretical Basis for Primary and Secondary Hydrogen Isotope Effects," in P.F. Cook (ed.) *Enzyme Mechanisms from Isotope Effects,* CRC Press, Boca Raton, (1991) pp. 3-35.

11. Klinman, J.P. "Hydrogen Tunneling and Coupled Motion in Enzyme Reactions," in P.F. Cook (ed.) *Enzyme Mechanisms from Isotope Effects,* CRC Press, Boca Raton (1991) pp. 127-48.

12. Schowen, K.B. and Schowen, R.L. "Solvent Isotope Effects on Enzyme Systems," *Methods Enzymol.* 87, 551-606 (1982).

13. Northrop, D.B. "Deuterium and Tritium Kinetic Isotope Effects on Initial Rates," *Methods Enzymol.* 87, 607-25 (1982).

14. Cook, P.F. "Kinetic and Regulatory Mechanisms of Enzymes from Isotope Effects," in P.F. Cook (ed.) *Enzyme Mechanisms from Isotope Effects,* CRC Press, Boca Raton (1991) pp. 203-30.

15. Weiss, P.M., Gavva, S.R., Harris, B.G., Urbauer, J.L., Cleland, W.W. and Cook, P.F. "Multiple Isotope Effects with Alternative Dinucleotide Substrates as a Probe of the Malic Enzyme Reaction," *Biochemistry,* 30, 5755-63.

16. DeJuan, E. and Taylor, K.B. "Isotope Effects in the Binding of NADH to Equine Liver Alcohol Dehydrogenase," *Biochemistry,* 15, 2523-2527 (1976).

17. Cleland, W.W. "Determination of Equilibrium Isotope Effects by the Equilibrium Perturbation Method," *Methods Enzymol.* 87, 641-6 (1982).

18. Swain, C.G., Silvers, E.C., Reuwer, J.F., Jr. and Schaad, L.J. "Use of Hydrogen Isotope Effects to Identify the Attacking Nucleophile in the Enolization of Ketones

Catalyzed by Acetic Acid, *J. Am. Chem. Soc.* 80, 5885-93 (1958).

19. Northrop, D.B. "Determining the Absolute Magnitude of Hydrogen Isotope Effects," in W.W. Cleland, M.H. O'Leary and D.B. Northrop (eds.) *Isotope Effects on Enzyme-Catalyzed Reactions*, University Park Press, Baltimore, pp 122-152.

20. Schimmerlik, M.I., Grimshaw, C.E. and Cleland, W.W. "Determination of the Rate-Limiting Steps for Malic Enzyme by the use of Isotope Effects and other Kinetic Studies ," *Biochemistry*, 16, 571-575.

21. Hermes, J.F., Roeske, C.A., O'leary, M.H. and Clelend, W.W. "Use of Multiple Isotope Effects to Determine Enzyme Mechanisms and Intrinsic Isotope Effects. Malic Enzyme and Glucose-6-Phosphate Dehydrogenase," *Biochemistry,* 21, 5106-5114 (1982).

22. Scheuring, J. and Schramm, V.L. "Kinetic Isotope Effect Characterization of the Transition State for Oxidized Nicotinamide Adenine Dinucleotide Hydrolysis by Pertussis Toxin," *Biochemistry*, 36, 4526-4534.

23. Schowen, K.B.J. "Solvent Hydrogen Isotope Effects," in R.D. Gandour and R.I. Schowen (eds.) *Transition States of Biochemical Processes,* Plenium Press, New York, (1978) pp. 225-80.

24. Quinn, D.M. and Sutton L.D. "Theoretical Basis and Mechanistic Utility of Solvent Isotope Effects," in P.F. Cook (ed.) *Enzyme Mechanism from Isotope Effects,* CRC Press, Boca Raton (1991) pp. 73-126.

12.6. Appendix 12.1: Derivation of Mathematical Model for the ^{13}C-Isotope Effects in the Presence and Absence of Deuterium

In the experiment to be modeled the ^{13}C-isotope effect is determined for the decarboxylation of malate catalyzed by malate dehydrogenase. The isotope effect is determined with and without ^{2}H in the oxidized position of malate. These two isotope effects are compared.

1. The mathematical model (equation 12.13) for the ^{13}C-isotope effects in the absence of ^{2}H is analogous to that for a 2H-isotope effect seen above (equation 12.4). In the present model (equation 12.12) the subscripts 12, 13, H, and D refer to isotopes.

$$\left[\frac{\left(\frac{V_{max}}{K_M}\right)_{12}}{\left(\frac{V_{max}}{K_M}\right)_{13}}\right]_H = \sum\left(\frac{k_{12}}{k_{13}}\right)*\Theta_{kH} + \sum\left(\frac{K_{12}}{K_{13}}\right)*\Theta_{KH} + \sum\Theta_H \tag{12.13}$$

2. The mathematical model (equation 12.14) for the ^{13}C-isotope effects in the presence of ^{2}H is similar, except that determinancy coefficients have a different meaning.

$$\left[\frac{\left(\frac{V_{max}}{K_M}\right)_{12}}{\left(\frac{V_{max}}{K_M}\right)_{13}}\right]_H = \sum\left(\frac{k_{12}}{k_{13}}\right)*\Theta_{kD} + \sum\left(\frac{K_{12}}{K_{13}}\right)*\Theta_{KD} + \sum\Theta_D \tag{12.14}$$

3. For example the determinancy coeffecient for those transitions that have the a ^{13}C-rate isotope effect in the absence of ^{2}H is:

$$\Theta_{kH} = \frac{\sum\frac{1}{k_{13}}*\prod\frac{1}{K}}{\sum\frac{1}{k_{13}}*\prod\frac{1}{K} + \sum\frac{1}{k}*\prod\frac{1}{K_{13}} + \sum\frac{1}{k}*\prod\frac{1}{K}} \tag{12.15}$$

where k_{13} and K_{13} denote those transitions that have a 13C-isotope effect in the rate or in the equilibrium respectively.

4. The determinancy coefficient in equation 12.14 is a function of the analogous coefficient in equation 12.13, and the analogous intrinsic ^{2}H-isotope effect, rate or equilibrium, on the V_{max}/K_M. The relationship depends upon the relative order of dehydrogenation and decarboxylation in the reaction sequence.

5. If the two processes are concurrent, the same transitions are affected by both isotope effects. The numerator of the determinancy coefficient for the rate isotope effects will be multiplied by the ^{2}H-rate isotope effect as will the same term in the denominator. For example:

$$\Theta_{kD} = \frac{\left(\dfrac{k_H}{k_D}\right) * \sum \dfrac{1}{k_{13}} * \prod \dfrac{1}{K}}{\left(\dfrac{k_H}{k_D}\right) * \sum \dfrac{1}{k_{13}} * \prod \dfrac{1}{K} + \left(\dfrac{K_H}{K_D}\right) * \sum \dfrac{1}{k} * \prod \dfrac{1}{K_{13}} + \sum \dfrac{1}{k} * \prod \dfrac{1}{K}} \qquad (12.16)$$

6. The numerator of the determinancy coefficient for the equilibrium isotope effects will be multiplied by the ^{2}H-rate isotope effect as will the same term in the denominator.

7. Since both of these determinancy coefficients are increased, the measured isotope effect will be greater than that in the absence of deuterium. The numerator terms for he other two determinancy coefficients can be seen in the denominator term, one with $1/K_{13}$ and one with no subscripts. The denominator for both of these will be the same as that in equation 12.16.

8. If dehydrogenation precedes decarboxylation, all of the transitions that have a ^{2}H-rate isotope effect will have no ^{13}C-isotope effect. Therefore, the ^{2}H-isotope effect will augment those determinancy coefficients that have no ^{13}C-isotope effects. Thus:

$$\Theta_D = \frac{\left(\dfrac{k_H}{k_D}\right) * \sum \dfrac{1}{k_H} * \prod \dfrac{1}{K} + \sum \dfrac{1}{k} * \prod \dfrac{1}{K}}{D}$$

$$D = \left(\dfrac{K_H}{K_D}\right) * \sum \dfrac{1}{k_{13}} * \prod \dfrac{1}{K} + \left(\dfrac{K_H}{K_D}\right) * \sum \dfrac{1}{k} * \prod \dfrac{1}{K_{13}} +$$

$$\left(\dfrac{k_H}{k_D}\right) * \sum \dfrac{1}{k_H} * \prod \dfrac{1}{K} + \sum \dfrac{1}{k} * \prod \dfrac{1}{K}$$

(12.17)

Since $\sum \Theta = 1.0$, the coefficients for the ^{13}C-isotope effect must decrease. Therefore, the measured ^{13}C-isotope effect will be less than that in the absence of ^2H.

9. The numerator expressions for the other determinancy coefficients can be seen in the expression for the denominator, one with $1/k_{13}$, one with $1/K_{13}$. The coefficient for the ^{13}C-rate and that for the ^{13}C-equilibrium isotope effect will be augmented in similar fashion by the ^2H-equilibrium isotope effect, but that effect is rather modest, 1.18.

10. If decarboxylation precedes dehydrogenation, none of the transitions affected by the ^{13}C-rate isotope effect will be affected by either the ^2H-rate isotope effect or the ^2H-equilibrium isotope effect. Although, some of the transitions affected by the ^{13}C-equilibrium isotope effect will be augmented by the ^2H-rate isotope effect, the former cannot exceed the intrinsic equilibrium effect, which is quite small. Therefore, the affect of ^2H on the measured ^{13}C-isotope effect is little or none. For example:

$$\Theta_{kD} = \frac{\sum \dfrac{1}{k_{13}} * \prod \dfrac{1}{K}}{D}$$

$$D = \sum \dfrac{1}{k_{13}} * \prod \dfrac{1}{K} + \left(\dfrac{k_H}{k_D}\right) * \sum \dfrac{1}{k} * \prod \dfrac{1}{K_{13}} + \sum \dfrac{1}{k} * \prod \dfrac{1}{K_{13}} +$$

$$\sum \dfrac{1}{k} * \prod \dfrac{1}{K}$$

(12.18)

CHAPTER 13

EFFECTS OF OTHER REACTION CONDITIONS

13.1 Introduction

In the previous two chapters the mathematical models for the effects of pH and the effects of isotopic substitution on the ratio of rates and equilibria of a single-step reaction at some standard set of conditions to that under the same set of conditions except the one condition that is perturbed were substituted into the general mathematic model for the ratio of initial velocities. In the present chapter this approach will be extended to the effects of temperature, pressure and chemical substitution (substituent effects). In addition the approach will be extended to the effects of simultaneous changes in more than one environmental condition.

13.1.1. OBJECTIVES

The objectives of this chapter are to show how informative experiments might be performed, how the experimental data can be interpreted in terms of mathematical models and an approach to the definition of the limitations of those interpretations. Efforts will also be made to describe the data from studies already published.

Since there is a relative paucity of investigations of the latter effects on initial velocity, the present chapter is somewhat more speculative in nature. Furthermore, the definition of the circumstances under which these investigations can produce useful mechanistic information and the data requirements are not known with certainty.

13.2 Effects of Temperature

Although there are a number of investigations of the effects of temperature on the initial velocity of enzymatic reactions [1], there are no systematic interpretations of the data in terms of general chemical and mathematical models of enzymatic reactions.

13.2.1. EXPERIMENTS

The experimental objective would be to determine the initial velocity and its operational parameters at a number of different temperatures, and would ordinarily be accomplished by the same techniques described earlier (Chapter 2). The techniques for the investigation of the effects of temperature on initial velocity are rather straight forward in the range close to ambient temperature (± 15 °C). However, outside this range problems arise in the maintenance of precise temperature. In addition to water-jacketed reactors (*e.g.* cuvettes), the precise maintenance of temperature during the transfer of significant volumes of solution to the reaction mixture has required the use of water-jacketed pipettes [2]. It is also very useful to have a precise temperature sensor in the reaction mixture itself.

The investigator must be sensitive to the fact that the kinetically essential pK values of one or more enzyme forms may change significantly with temperature. Therefore, it is necessary to conduct the experiments at a region of the pH profile where the change with pH is minimal and confirm that such is true at each temperature. Therefore, it may be necessary to investigate the effects of pH at several different temperatures.

As discussed somewhat later the maximum precision of data over the longest temperature range possible would be desirable. Therefore, the temperatures approaching 0 °C are as useful as those higher than ambient. In addition it may be desirable to perform all of the experiments in the presence of solvents that lower the freezing point significantly (e.g. ethylene glycol or glycerol) in order to extend the range of possible experiments.

13.2.2. MODELS AND DATA

$$\frac{K_0}{K_e} = \exp\left[\left(\frac{\Delta H}{R}\right) * \left(\frac{1}{T_0} - \frac{1}{T_e}\right)\right]$$

$$\frac{k_0}{k_e} = \frac{T_0}{T_e} * \exp\left[\left(\frac{-\Delta H^\ddagger}{R}\right) * \left(\frac{1}{T_0} - \frac{1}{T_e}\right)\right]$$

(13.1)

The mathematical model for the effect of a change in temperature on an equilibrium and on a rate respectively (equation 13.1) is a function of the enthalpy, ΔH, of the equilibrium and of the enthalpy of activation, $\Delta H^\ddagger$, of the rate constant respectively. Substitution of these models into the mathematical model for the ratio of initial velocities of a chemical model (Figure 13.1) for a unireactant enzyme with one enzyme-substrate complex and one enzyme-product complex under two sets of environmental conditions (equation 10.24) results in a

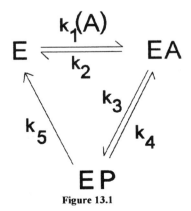

Figure 13.1

mathematical model for the effect of a change in temperature from T_0 to T_e, on initial velocity (equation 13.2). The overall form of the model is a sum of exponential terms, one for each transition. Each of the terms is multiplied by the determinancy coefficient, Θ, for that transition.

$$
\begin{aligned}
\frac{v_{i0}}{v_{ie}} = {} & \Theta_1 * \frac{T_0}{T_e} * \exp\left[\left(\frac{-\Delta H_1^{\ddagger}}{R}\right)*\left(\frac{1}{T_0}-\frac{1}{T_e}\right)\right] + \Theta_{13} * \frac{T_0}{T_e} * \exp\left[\left(\frac{-\Delta H_{13}^{\ddagger}}{R}\right)*\left(\frac{1}{T_0}-\frac{1}{T_e}\right)\right] + \\
& \Theta_{15} * \frac{T_0}{T_e} * \exp\left[\left(\frac{-\Delta H_{15}^{\ddagger}}{R}\right)*\left(\frac{1}{T_0}-\frac{1}{T_e}\right)\right] + \Theta_3 * \frac{T_0}{T_e} * \exp\left[\left(\frac{-\Delta H_3^{\ddagger}}{R}\right)*\left(\frac{1}{T_0}-\frac{1}{T_e}\right)\right] + \\
& \Theta_{35} * \frac{T_0}{T_e} * \exp\left[\left(\frac{-\Delta H_{35}^{\ddagger}}{R}\right)*\left(\frac{1}{T_0}-\frac{1}{T_e}\right)\right] + \Theta_5 * \frac{T_0}{T_e} * \exp\left[\left(\frac{-\Delta H_5^{\ddagger}}{R}\right)*\left(\frac{1}{T_0}-\frac{1}{T_e}\right)\right]
\end{aligned} \tag{13.2}
$$

The mathematical models for k_{cat}/K_M and for k_{cat} will have a similar format (equation 13.3 and equation 13.4).

$$
\begin{aligned}
\frac{\left(\dfrac{k_{cat}}{K_A}\right)_0}{\left(\dfrac{k_{cat}}{K_A}\right)_e} = {} & \Theta_1 * \frac{T_0}{T_e} * \exp\left[\left(\frac{-\Delta H_1^{\ddagger}}{R}\right)*\left(\frac{1}{T_0}-\frac{1}{T_e}\right)\right] + \\
& \Theta_{13} * \frac{T_0}{T_e} * \exp\left[\left(\frac{-\Delta H_{13}^{\ddagger}}{R}\right)*\left(\frac{1}{T_0}-\frac{1}{T_e}\right)\right] + \Theta_{15} * \frac{T_0}{T_e} * \exp\left[\left(\frac{-\Delta H_{15}^{\ddagger}}{R}\right)*\left(\frac{1}{T_0}-\frac{1}{T_e}\right)\right]
\end{aligned} \tag{13.3}
$$

$$
\begin{aligned}
\frac{\left(k_{cat}\right)_0}{\left(k_{cat}\right)_e} = {} & \Theta_3 * \frac{T_0}{T_e} * \exp\left[\left(\frac{-\Delta H_3^{\ddagger}}{R}\right)*\left(\frac{1}{T_0}-\frac{1}{T_e}\right)\right] + \Theta_{35} * \frac{T_0}{T_e} * \exp\left[\left(\frac{-\Delta H_{35}^{\ddagger}}{R}\right)*\left(\frac{1}{T_0}-\frac{1}{T_e}\right)\right] + \\
& \Theta_5 * \frac{T_0}{T_e} * \exp\left[\left(\frac{-\Delta H_5^{\ddagger}}{R}\right)*\left(\frac{1}{T_0}-\frac{1}{T_e}\right)\right]
\end{aligned} \tag{13.4}
$$

Models for the interpretation of temperature effects must include the possibility that some temperatures might enlarge the number of alternative conformations of enzyme intermediates in equilibrium with those at the standard temperature and that one or more of the new conformations may react at a different rate, or not react at all. Although irreversible conformation changes (denaturation) and slow, reversible conformation changes can be identified from the results of time-course experiments rapid reversible conformation changes can be somewhat more subtle.

The latter may be modeled as an alternative enzyme form in equilibrium with an enzyme form already a member of the reaction sequence (*e.g.* Figure 13.2). Although the new conformation, E_3', may not actually be in rapid equilibrium with the catalytic enzyme form, E_3, it can be modeled as such since it does not react further. The mathematical models that include such equilibria can be developed in similar manner to those for pH effects (Chapter 11). A conformation change affecting only one or more intermediates in a transition will have no affect

Figure 13.2

on the overall transition, since it will affect both the forward and reverse reactions from that intermediate and, thus, cancel out of the mathematical model for that transition. However, a reversible conformation change of the reactant enzyme form for a given transition will affect the mathematical model for that transition (equation 13.5).

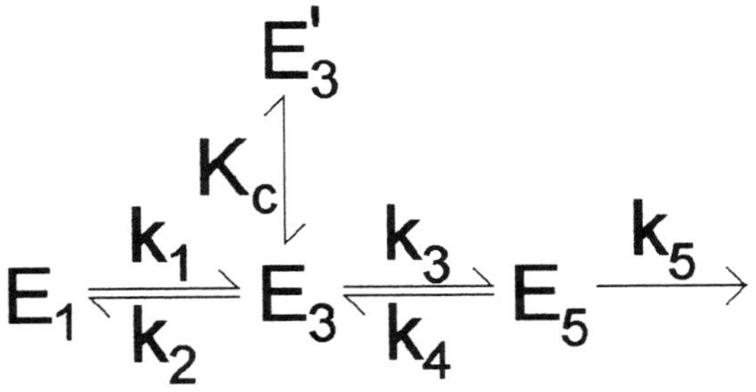

$$\frac{[K_1*K_3*k_5]_0}{[K_1*K_3*k_5]_e} = \frac{T_0}{T_T}*\exp\left[\frac{-\Delta H_{15}^{\ddagger}}{R}*\left(\frac{1}{T_0}-\frac{1}{T_T}\right)\right]*\left[1+K_c*\exp\left[\frac{-\Delta H_c}{R}*\left(\frac{1}{T_0}-\frac{1}{T_T}\right)\right]\right] \quad (13.5)$$

If that transition is substantially rate determining for an experimental parameter, its mathematical model will be the sum of two exponential terms, one reflecting the enthalpy for the overall transition and the other reflecting the sum of that and the enthalpy for the conformation change, ΔH_c.

The models for the effect of temperature on initial velocity and its operational parameters are all sums of exponential terms and the ratio of the parameters at the two temperatures will be determined by the significantly rate-determining energy transitions. Therefore, it should theoretically be possible to determine the number of transitions that are significantly rate determining as well as the relative extent to which each is rate determining (*i.e* the determinancy coefficient) and the enthalpy of activation of each such transition. There are a number of techniques for exponential curve fitting [3] to determine the number of significantly rate-determining steps as well as the enthalpy of activation of each. However, it should be simple to fit the data to models with the same format but with different numbers of exponential terms. The model with the lowest number of exponential terms that can be distinguished from all those models with even fewer terms is the best model to fit the data.

From the value for each of the determinancy coefficients in the simple model with no adventitious conformation changes it is possible to calculate the $\Delta G^{\ddagger}$ for each of the significantly rate-determining transitions and consequently, the $\Delta S^{\ddagger}$ for each transition. For example the expression for one of the determinancy coefficients in the k_{cat}/K_A for the model above is:

$$\Theta_1 = \frac{\dfrac{1}{k_1}}{\dfrac{1}{k_1} + \dfrac{1}{K_1 * k_3} + \dfrac{1}{K_1 * K_3 * k_5}}$$

$$= \left(\frac{k_{cat}}{K_A}\right)_0 * \frac{h}{k_b * T} * \exp\left[\frac{\Delta G_1^{\ddagger}}{R * T}\right]$$

(13.6)

However, it must be realized that one or more of the processes may represent conformation changes in one or more enzyme forms.

The success of the interpretation of temperature effects would depend on the range of temperatures investigated, the precision of the data, and the relative magnitude of the differences in their enthalpy values. A longer range of experimental temperatures and more precise data will increase the possibility for detection of more than one rate-determining transition. Of course more than one rate-determining transition with equal or similar enthalpy values will be indistinguishable. Elongation of the temperature range might be aided by investigations of an enzyme from a thermophilic organism and/or use of solvents, described above, to extend the range below freezing. Of course in the latter approach the same solvents must be used in the experiments above freezing. An investigation of the required limits of the precision of the data, the temperature range and the differences in enthalpy for the detection of more than one rate-determining transition of as well as of their detailed interrelationships would seem to be useful for possible future experimental investigations.

Although the number of significantly rate-determining transitions and their respective thermodynamic quantities might be calculated, the interpretation of the latter and the identification of the chemical steps involved remains problematic. Thus with a given set of experimental data it might be difficult to distinguish between a model with a change in the rate-determining step and one with an adventitious conformation change, even if there were significant differences in the entropy.

13.3. Substituent Effects

Although the prediction of the effects of the substitution of various chemical groups near the reaction center on nonenzymatic reaction has enjoyed some success, the analogous efforts with enzyme-catalyzed reactions would seem to be fraught with a bewildering array of difficulties. In fact the prediction of the effects of more or less electron withdrawing or electron donating groups on the initial velocity and its operational parameters is associated with such uncertainties that the systematic study of these effects has seldom been reported. Among the notable exceptions are those by Klinman [4] on alcohol dehydrogenase and dopamine-β-hydroxylase [5].

In the present section an approach will be developed to interpret data from experiments in which the initial velocity and its operational parameters are measured for a series of substrates in which various chemical groups have been substituted. The objective is to develop an approach by which hypotheses about the nature of the transitions states in the reaction may be accepted or rejected. However, in the absence of success with this objective it should be possible to define the sources of the uncertainties and the limitations of the approach.

13.3.1. THE EXPERIMENTS

The experimental objective would be to determine the initial velocity and its operational parameters with a number of different substrates. It would ordinarily be accomplished by the same techniques described earlier (Chapter 2). However, it may also be useful to perform competition experiments similar to those for the determination of isotope effects (Chapter 12). Of course direct comparison experiments must be done under the same conditions except for the substrate, and the investigator must be sensitive to the possibilities that any of the experimental substrates may differ in properties that affect the reaction in more obvious ways, e.g. solubility, pK, etc.

13.3.2. MODELS AND DATA

The effect of inductive and conjugative effects of neighboring substituents on the rate and equilibrium of nonenzymatic organic chemical reactions have been studied for years in attempts to quantitate, explain and predict them with meaningful chemical and mathematical models. Although there are a number of related mathematical models known as linear-free-energy relationships, the most successful has been that of the electronic effect of substituents on aromatic rings [6]. In the simplest model (equation 13.7) the ratio of the rate or equilibrium constant of the reaction with a standard compound, usually with a hydrogen atom (K_H or k_H) in the appropriate position to that of some appropriate substituted compound (K_S or k_S) is related to the exponential of the product of a factor, σ and a factor, ρ. The original definitions of σ and ρ were for the dissociation constants of a series of substituted benzioc acids.

$$\frac{K_H}{K_S} = \exp[\rho * \sigma]$$

$$\frac{k_H}{k_S} = \exp[\rho * \sigma]$$

(13.7)

The factor, σ_a, is a measure of the electron withdrawing power of the group by inductive or resonance effect on the Sn-1 substitution reaction of dimethylbenzylchloride [7]. Therefore, factor, ρ, is a measure of the positive charge built up on the reactive center in the transition state or the product (in the case of an equilibrium). The literature contains tables of σ_a and ρ factors for a large number of substituent groups and organic reactions [7]. Somewhat less successful has been the prediction of the effects of substituent groups on the equilibria and reaction rates of aliphatic compounds, but a similar mathematical model has been utilized with somewhat different sets of values for ρ and σ [8]. Furthermore, there are values, for factors analogous to ρ and σ values, for hydrophobicity [9] and for size [10] of the substituent groups.

Substitution of the chemical model for electron withdrawal on single aromatic rings into the mathematical models of environmental effects on initial velocity (equation 10.24) and those for its operational parameters (equations 10.29 and 10.30) for a unireactant model with one enzyme-substrate and one enzyme-product intermediate (Figure 13.1) results in mathematical models for substituent effects (equation 13.8, equation 13.9). The hypothetical ρ values are those for the transition from the intermediate to the transition state on the chemical pathway catalyzed by the enzyme and may involve several chemical steps.

$$\frac{v_{iH}}{v_{iS}} = \Theta_1 * \exp[\rho * \sigma] + \Theta_{13} * \exp[\rho * \sigma] + \Theta_{15} * \exp[\rho * \sigma] +$$

$$\Theta_3 * \exp[\rho * \sigma] + \Theta_{35} * \exp[\rho * \sigma] + \Theta_5 * \exp[\rho * \sigma]$$

(13.8)

$$\frac{\left[\dfrac{k_{cat}}{K_A}\right]_H}{\left[\dfrac{k_{cat}}{K_A}\right]_S} = \Theta_1 * \exp[\rho * \sigma] + \Theta_{13} * \exp[\rho * \sigma] + \Theta_{15} * \exp[\rho * \sigma]$$

(13.9)

$$\frac{[k_{cat}]_H}{[k_{cat}]_S} = \Theta_3 * \exp[\rho * \sigma] + \Theta_{35} * \exp[\rho * \sigma] + \Theta_5 * \exp[\rho * \sigma]$$

The mathematical model has the form of a sum of exponential terms each corresponding to a different energy transition and multiplied by the corresponding determinancy coefficient, Θ, similar to that for temperature effects. Therefore, it is theoretically possible to determine the number of rate determining transitions that have different values for ρ, and the determinancy coefficient for each. The strategy for the identification of the best model will be the same as that for the temperature effects. If a dependable ρ-value could be determined, it should be possible to eliminate certain hypotheses and support others about the chemical nature of the rate-determining transition states.

In several respects the possibilities for success in the interpretation of substituent effects would seem to be better than that of temperature effects. The experiments should be less laborious, since it is usually not necessary to consider the effects of pH with each substituent. In addition peripheral phenomena such as conformation changes would ordinarily not be different with different substrates. Furthermore, since the range of possible values of σ, for substituents on aromatic rings at least, is greater than one $\log_{10}$, whereas that for a feasible temperature investigation is generally less. However, other uncertainties associated with the investigation of substituent effects limits its current usefulness.

Of the other properties of substituent groups that might affect the rate he most obvious example is that of the bulk properties. The hydrated size of some substituent groups may limit its ability to fit in the active site of the enzyme. Nevertheless it is also possible to interpret the data with analogous sets of values for σ, including one based on size, σ_d, and one based on hydrophobicity, σ_h, [9]. A common mathematical model for these several dimensions of effects is to replace the simple product of σ and ρ with a sum of products, each for a different dimension of effect (*e.g.* equation 13.10). However, it is uncertain whether this more complex multidimension equation in the exponent could be distinguished from a sum of exponents each with a single dimension (*e.g.* equation 13.9).

$$\frac{\left[\dfrac{k_{cat}}{K_A}\right]_H}{\left[\dfrac{k_{cat}}{K_A}\right]_S}=\Theta_1*\exp\left[\rho_a*\sigma_a+\rho_d*\sigma_d+\rho_h*\sigma_h\right]_1+\Theta_{13}*\exp\left[\rho*\sigma_a+\rho_d*\sigma_d+\rho_h*\sigma_h\right]_{13}+$$

(13.10)

$$\Theta_{15}*\exp\left[\rho*\sigma_a+\rho_d*\sigma_d+\rho_h*\sigma_h\right]_{15}$$

For example in the investigation of the effects of substituents on the oxidation of benzylic alcohol by alcohol dehydrogenase from yeast Klinman [4] fit data for the log of the parameter by three dimensional linear regression to the three-dimensional mathematical model and found that most parameters were affected in only one or two dimensions. For example the catalytic constant (k_{cat}) for the reduction of benzaldehydes was sensitive only to electron withdrawing groups, whereas that for the oxidation of benzyl alcohol was insensitive to all three effects. The binding of benzyl alcohol was promoted by hydrophobic groups. Although no formal attempt was made to identify the rate-determining transitions, the substantial primary deuterium isotope effect on the reduction of benzyl alcohols indicates that catalysis is rate determining to a significant extent for the k_{cat}.

However, other sources of uncertainty include differences in hydrogen bonding and ionic attractions in the active site of the enzyme that would cause certain values for the initial velocity and its operational parameters to deviate from the remainder of the series. The possibilities of success of the investigation might be enhanced if the substrates were limited to a homologous series such as alkyl or fluorinated derivatives. The latter has the possible advantage that the size is minimal.

In summary although the interpretation of the results of investigations of substituent effects on the initial velocity of enzyme catalyzed reactions and its operational parameters to learn mechanistic information would seem to be fraught with uncertainties, models are proposed that should provide a possible approach.

13.4. The Effects of Pressure

Although the effect of pressure on initial velocity and its operational parameters is conceptually both interesting and interpretable, in principal, with the mathematical models presented here, the experimental and technical problems are formidable. Nevertheless Northrup [11] has reported the pressure effects on a number of dehydrogenase enzymes, and interpreted the results in the context of reverse and forward commitment factors.

13.4.1. THE EXPERIMENTS

Initial velocity experiments can be conducted and data can be collected with commercially available equipment [11]. However, because significant time may be required for constitution and pressurization of the reaction mixture it may be necessary to extrapolate the collected data to calculate initial velocity.

The investigator must be sensitive to the effect of pressure on temperature and on the pK of the buffer employed. The former can be controlled by strict attention to the actual temperature of the reaction mixture and the use of a temperature-controlled reaction vessel, *e.g.* cuvette; whereas the latter can be controlled by the selection of a buffer whose pK is relatively insensitive to pressure.

Somewhat more difficult to control is the possibility that the pK of essential groups on the enzyme may change with pressure. For this reason it is advisable for the investigator to perform the measurements at more than one proximate pH value to provide assurance that the change of the initial velocity due to a shift in pK is minimal at the reaction pH.

13.4.2. MODELS AND DATA

The effect of pressure on the equilibrium and rate constants of a single-step nonenzymatic reaction depends on the volume change between the reactants and either the products or the transition state respectively (equation 13.11).

$$\frac{K_0}{K_P} = \exp\left[\left(\frac{\Delta V}{R*T}\right)*(P_P - P_0)\right]$$

$$\frac{k_0}{k_P} = \exp\left[\left(\frac{\Delta V^{\ddagger}}{R*T}\right)*(P_P - P_0)\right]$$

(13.11)

Substitution of these mathematical models into the general mathematical models of environmental effects on initial velocity (equation 10.24) and those for its operational parameters (equations 10.29 and 10.30) for a unireactant model with one enzyme-substrate intermediate and one enzyme-product intermediate (Figure 13.1) results in mathematical models for the effect of pressure on the latter (equation 13.12, equation 13.13).

$$\frac{v_{i0}}{v_{ie}} = \Theta_1 * \exp\left[\left(\frac{\Delta V_1^{\ddagger}}{R*T}\right) * (P_e - P_0)\right] + \Theta_{13} * \exp\left[\left(\frac{\Delta V_{13}^{\ddagger}}{R*T}\right) * (P_e - P_0)\right] +$$

$$\Theta_{15} * \exp\left[\left(\frac{\Delta V_{15}^{\ddagger}}{R*T}\right) * (P_e - P_0)\right] + \Theta_3 * \exp\left[\left(\frac{\Delta V_3^{\ddagger}}{R*T}\right) * (P_e - P_0)\right] +$$

$$\Theta_{35} * \exp\left[\left(\frac{\Delta V_{35}^{\ddagger}}{R*T}\right) * (P_e - P_0)\right] + \Theta_5 * \exp\left[\left(\frac{\Delta V_5^{\ddagger}}{R*T}\right) * (P_e - P_0)\right]$$

(13.12)

$$\frac{\left[\dfrac{k_{cat}}{K_A}\right]_0}{\left[\dfrac{k_{cat}}{K_A}\right]_e} = \Theta_1 * \exp\left[\left(\frac{\Delta V_1^{\ddagger}}{R*T}\right) * (P_e - P_0)\right] + \Theta_{13} * \exp\left[\left(\frac{\Delta V_{13}^{\ddagger}}{R*T}\right) * (P_e - P_0)\right] +$$

$$\Theta_{15} * \exp\left[\left(\frac{\Delta V_{15}^{\ddagger}}{R*T}\right) * (P_e - P_0)\right]$$

(13.13)

$$\frac{[k_{cat}]_0}{[k_{cat}]_e} = \Theta_3 * \exp\left[\left(\frac{\Delta V_3^{\ddagger}}{R*T}\right) * (P_e - P_0)\right] + \Theta_{35} * \exp\left[\left(\frac{\Delta V_{35}^{\ddagger}}{R*T}\right) * (P_e - P_0)\right] +$$

$$\Theta_5 * \exp\left[\left(\frac{\Delta V_5^{\ddagger}}{R*T}\right) * (P_e - P_0)\right]$$

These mathematical models are sums of exponential terms similar to the model for temperature effects and that for substituent effects. Therefore, it should be theoretically possible to determine the approximate number of rate-determining transitions, the value of the determinancy coefficient for each, and the volume change associated with each of them. The latter might help identify the chemical change associated with these transitions. However, the limits of data precision, range of pressures investigated, and differences in the volume changes, ΔV's, required for useful evidence for more than one rate-determining transition are not known. In addition the interpretation may be complicated by a conformation change of one or more enzyme intermediates at different pressures. In fact it seems likely that a significant volume change would most likely be due to a conformation change, whether it occurred in concert with one of the binding and/or catalytic steps or adventitious to the catalytic progress of the reaction. Models for an adventitious conformation change can be developed as those described above for temperature effects. As in the case of temperature effects the mathematical models that include an inactive conformation change are indistinguishable from that for more than one rate-determining transition.

13.5. Other Environmental Effects

Conceptual chemical and mathematical models can be developed for other environmental effects, but further development of them is even more speculative than those above. For example a model for the effect of ionic strength might be developed on the basis of charge separation in the transition states.

13.6. Composite Environmental Effects

Although the investigation of the effects of a combination of environmental factors would seem to be an attractive option, the models for data interpretation are complicated even further by the possibility of secondary effects between factors. Nevertheless several investigations have been carried out on a combination of factors whose secondary effects would seem to be minimal.

13.6.1. ISOTOPE AND pH EFFECTS

For example isotope effects and pH effects have been determined together by a number of investigators [12] using a combination of the experimental techniques described previously.

The mathematical model for interpretation would be derived from the model corresponding to the example for general environmental effects (equation 10.24) in which ratio of the rate constants for the isotope effect on each transition (equation 12.3) would be multiplied by the pH-function for that transition. For example the mathematical model for the composite effects on the k_{cat}/K_A for the model illustrated above (Figure 13.1) with an essential basic group is equation 13.14.

$$\frac{\left[\dfrac{k_{cat}}{K_A}\right]_D}{\left[\dfrac{k_{cat}}{K_A}\right]_e} = \Theta_1 * \frac{k_1}{k_{1i}} * \frac{1}{1 + \dfrac{(H^+)}{K_{al}}} + \Theta_{13} * \frac{K_1 * k_3}{K_{1i} * k_{3i}} * \frac{1}{1 + \dfrac{(H^+)}{K_{al}}} +$$

$$\Theta_{15} * \frac{K_1 * K_3 * k_5}{K_{1i} * K_{3i} * k_{5i}} * \frac{1}{1 + \dfrac{(H^+)}{K_{al}}}$$

(13.14)

If the simple chemical model in which the only reactant for all of the transitions in the k_{cat}/K_A is the free enzyme, as in equation 13.14, is acceptable; and if the simple chemical model in which there is only one isotopically sensitive step is acceptable (equation 12.4); the measured primary isotope effect should not change with pH and the apparent pK should not change in the presence of isotope, since the pH function is the same for all of the terms in the model. However, if some of the transitions are affected differently by pH, both the isotope effect should change with pH and the apparent pK should change in the presence of isotope. Furthermore, it should be possible to eliminate some hypothetical models with respect to which of the transitions is associated with the aberrant pH effect.

In models in which transitions are affected differently by pH and isotopic substitution, if the most pH-sensitive transition and the isotopically sensitive transition are the same, the isotope effect will be larger at a less optimum pH unless the intrinsic isotope effect is already fully expressed. Indeed this is the case in several studies. For example the primary deuterium isotope effect on both the k_{cat} and the k_{cat}/K_M for yeast alcohol dehydrogenase increase below the pK of an essential basic group on some enzyme form or forms [13]. Furthermore, if the pH-sensitive transition and the isotope-sensitive transition are the same the apparent pK of an essential basic group should increase and that of an essential acidic group should decrease in the measurements performed in the presence of isotope.

The enzyme form responsible for the effects of pH is always upstream from the transition state responsible for an enhanced isotope effect. Therefore, in the example above the pH-sensitive intermediate must be upstream from the catalytic step. If the measured parameter has a decreased isotope effect at suboptimal pH, the two affect different transitions.

13.6.2. ISOTOPE AND SUBSTITUENT EFFECTS

Data from the combination of isotope and substituent effects on the substrate could be interpreted with somewhat similar models, although the serious uncertainties associated with substituent effects discussed above must be recognized here also. The mathematical model (*e.g.* equation 13.15) would be similar to that for general environmental effects (equation 10.24) in which ratio of the rate constants for the isotope effect (equation 12.4) would be multiplied by the corresponding exponential function for the substituent effect presented above.

$$\frac{\left[\dfrac{k_{cat}}{K_A}\right]_D}{\left[\dfrac{k_{cat}}{K_A}\right]_e} = \Theta_1 * \frac{k_1}{k_{1i}} * \exp[\rho*\sigma] + \Theta_{13} * \frac{K_1*k_3}{K_{1i}*k_{3i}} * \exp[\rho*\sigma] +$$

$$\Theta_{15} * \frac{K_1*K_3*k_5}{K_{1i}*K_{3i}*k_{5i}} * \exp[\rho*\sigma]$$

(13.15)

If a substituent inhibits the initial velocity, or one of its parameters, by inhibiting the isotopically sensitive transitions; the isotope effect will increase unless the intrinsic isotope effect is already expressed to a maximum extent. If a substituent inhibits the initial velocity, or one of its parameters, by inhibiting a transition other than the isotopically sensitive transition; the isotope effect will decrease.

If a substituent accelerates the initial velocity, or one of its parameters, by increasing the rate of one of the isotopically sensitive transitions; the isotope effect will decrease. If a substituent accelerates the initial velocity, or one of its parameters, by increasing the rate of a transition other than the isotopically sensitive ones; the isotope effect will increase.

If the substituent effect for some parameter can be interpreted in an unambiguous manner, it may be possible to determine the order in the reaction sequence of the transition states responsible for the substituent effect and the isotope effect by application of logic analogous to that applied in the interpretation of dual isotope effects.

13.6.3. ISOTOPE AND PRESSURE EFFECTS

The combined effects of pressure and isotopic substitution might be interpreted with a similar mathematical model (*e.g.* equation 13.16), although the complexities associated with the interpretation of pressure effects alone must be considered here also. The combination of isotope and pressure effects has been investigated by Northrop, *e.g.* [14].

$$\frac{\left(\dfrac{k_{cat}}{K_A}\right)_0}{\left(\dfrac{k_{cat}}{K_A}\right)_e} = \Theta_1 * \frac{k_1}{k_{1i}} * \exp\left[\left(\frac{\Delta V_1^{\ddagger}}{R*T}\right)*(P_P - P_0)\right] + \Theta_{13} * \frac{K_1 * k_3}{K_{1i} * k_{3i}} * \exp\left[\left(\frac{\Delta V_{13}^{\ddagger}}{R*T}\right)*(P_P - P_0)\right] +$$

$$\Theta_{15} * \frac{K_1 * K_3 * k_5}{K_{1i} * K_{3i} * k_{5i}} * \exp\left[\left(\frac{\Delta V_{15}^{\ddagger}}{R*T}\right)*(P_P - P_0)\right]$$

(13.16)

If there is a significant increase in the volume of the enzyme in one or more of the isotopically sensitive transition states, an increase in pressure will result in an increased observed isotope effect unless the intrinsic isotope effect is already fully expressed. If there is a significant increase in the volume of the enzyme form in one or more of the transition states that are not isotopically sensitive, an increase in pressure will result in a decreased observed isotope effect.

If there is a significant decrease in the volume of the enzyme in one or more of the isotopically sensitive transition states, an increase in pressure will result in an decreased observed isotope effect. If there is a significant decrease in the volume of the enzyme in one or more of the transition states that are not isotopically sensitive, an increase in pressure will result in an increased observed isotope effect.

Whether or not all of these increases and decreases in the isotope effect can be measured depends on the measurement technique and the relative magnitudes of the determinancy coefficients, Θ.

If the pressure effects alone on a parameter can be interpreted in an unambiguous manner, it may be possible to eliminate one or more hypotheses about the relative order in the reaction sequence of the transition states responsible for the pressure effects and the isotope effects by logic analogous to that used in the interpretation of the effects of dual isotopic substitution.

13.7. Summary

Mathematical models similar to those useful in the interpretation of pH-effects and isotope effects may be useful in the interpretation of the effects of temperature, substituents, pressure and other environmental factors, as well as combinations of environmental factors. However, systematic experimental investigations of these effects are generally not plentiful and the models remain to be proven useful. Nevertheless the successful application of the models promises to reward the investigator with significant insight into the chemical mechanism of enzymatic reactions.

13.8. References

1. Laidler, K.F. and Peterman, B.F. "Temperature effects in Enzyme Kinetics," *Methods Enzymol.* 63, 234-57 (1979).

2. Ford, J.B., Askins, J. and Taylor, K.B. "Kinetic Models for Synthesis by a Thermophilic Alcohol Dehydrogenase," *Biotechnol. Bioeng.* 42, 367-75.

3. Halvorson, H.R. "Padé-Laplace Algorithm for Sums of Exponentials: Selecting Appropriate Exponential Model and Initial Estimates for Exponential Fitting," *Methods Enzymol.* 210, 54-67.

4. Klinman, J.P. "Isotope Effects and Structure-Reactivity Correlations in the Yeast Alcohol Dehydrogenase Reaction. A Study of the Enzyme-Catalyzed Oxidation of Aromatic Alcohols," *Biochemistry,* 15, 2018-2026.

5. Miller, S.M. and. Klinman, J.P. "Secondary Isotope Effects and Structure-Reactivity Correlations in the Dopamine-Beta-Monooxygenase Reaction: Evidence for a Chemical Mechanism," *Biochemistry.* 24, 2114-27 (1985).

6. Carroll, F.A. *Perspectives on Structure and Mechanism in Organic Chemistry,* Brooks/Cole Pub. Co., Boston, pp 366-385 (1998).

7. Hoefnagel, A.J. and Wepster, B.M. "Sumstituent Effects. IV. Reexamination of σ^n, $\Delta\sigma_R^+$, and σ_R^n Values; Arylacetic Acids and Other Insulated Systems," *J. Am. Chem. Soc.* 95, 5357-66 (1973).

8. Taft, R.W., Jr., "The Heneral Nature of the Proportionality of Polar Effects of Substituent Groups in Organic Chemistry," *J. Am. Chem. Soc.* 75, 4231-8 (1953).

9. Leo, A., Hansch, C. and Elkins, C. "Partition Coefficients and their Uses," *Chem. Rev.* 71, 525-616 (1971).

10. Pauling, L. *Nature of the Chemical Bond,* Cornell Univ. Press, Ithaca, p. 187 (1945).

11. Cho, Y.-K. and Northrop, D.B. "Effects of Pressure on the Kinetics of Capture by Yeast Alcohol Dehydrogenase," *Biochemistry,* 38, 7470-7475 (1999).

12. Cook, P.F. "pH Dependence of Isotope Effects in Enzyme Catalyzed Reactions," in P.F. Cook (ed.) *Enzyme Mechanism from Isotope Effects,* CRC Press, Boca Raton,

(1991) pp. 231-245.

13. Cook, P.F. and Cleland, W.W. "pH Variation of Isotope Effects in Enzyme-Catalyzed Reactions. I. Isotope- and pH-Dependent Steps the Same," *Biochemistry*, 20, 1797-805 (1981).

14. Quirk, D.J. and Northrop, D.B. "Effect of Pressure on Deuterium Isotope Effects of Formate Dehydrogenase," *Biochemistry*, 40, 847-51 (2001).

INDEX

223